D1485182

Valuing Climate Change

THE ECONOMICS OF THE GREENHOUSE

Samuel Fankhauser

WITHDRAWN FROM STOCK

E·S·R·C
ECONOMIC & SOCIAL RESEARCH COUNCIL

CENTRE FOR SOCIAL AND ECONOMIC RESEARCH ON THE GLOBAL ENVIRONMENT

C|S|E|R|G|E

5 060 744 8

KEELE UNIVERSITY
LIBRARY

0 9 MAY 1995

OL/281

First published in 1995 by
Earthscan Publications Limited
120 Pentonville Road, London N1 9JN

Copyright © Samuel Fankhauser, 1995

All rights reserved. No part of this book may be reproduced in any form without written permission from the publisher and author.

A catalogue record for this book is available from the British Library

ISBN: 1 85383 237 5

Typeset by DP Photosetting, Aylesbury, Bucks
Printed and bound by Clays Ltd, St Ives plc

Earthscan Publications Limited is an editorially independent subsidiary of Kogan Page Limited, and published in association with the International Institute for Environment and Development and the World Wide Fund for Nature.

Contents

List of Illustrations

FIGURES

BOXES

TABLES

List of Acronyms and Abbreviations

AEEI	Autonomous Energy Efficiency Improvements
BAU	Business as Usual
CETA	Carbon Emission Trajectory Assessment
CGE	Computable General Equilibrium
CRRA	Constant Relative Rate of Risk Aversion
CRTM	Carbon Rights Trade Model
DICE	Dynamic Integrated Climate Economy
EPA	US Environmental Protection Agency
FAO	Food and Agricultural Organisation
GCM	Global Circulation Models
GCAM	Global Change Assessment Model
GEF	Global Environment Facility
GDP	Gross Domestic Product
GNP	Gross National Product
GREEN	GeneRal Equilibrium ENvironmental
GWP	Global Warming Potential
HERMES	Harmonized Economic Research Models on Energy Systems
IEA	International Energy Agency
ICAM	Integrated Climate Assessment Model
IMAGE	Integrated Model for the Assessment of the Greenhouse Effect
IPCC	Intergovernmental Panel on Climate Change
MDM	Multisectoral Dynamic Model
MERGE	Model for Evaluating Regional and Global Effects
MINK	Missouri, Iowa, Nebraska, Kansas
MIT	Massachusetts Institute of Technology
OECD	Organization for Economic Cooperation and Development
PAGE	Policy Assessment of the Greenhouse Effect
PPM	Parts per million by volume
SGM	Second Generation Model
SLR	Sea Level Rise
UNEP	United Nations Environmental Programme
WHO	World Health Organisation

WRI   World Resources Institute

CHEMICAL SYMBOLS

C   Carbon
CFC   Chlorofluorocarbon
CO   Carbon monoxide
CO_2 Carbon dioxide
CH_4 Methane
HC   Hydrocarbon
HCFC   Hydrochlorofluorocarbon
N_2O Nitrous oxide
NO_x Nitrogen oxides
O_3 Ozone
SO_2 Sulphur dioxide
SO_x Sulphur oxides
TSP   Total Suspended Particles
VOC   Volatile Organic Compound

Foreword

by David W Pearce

Climate modellers tell us they cannot be certain that there is a long run trend of 'global warming', a rise in the mean surface temperature of the Earth. We must wait for better evidence, and the period of waiting could be ten years or so. This uncertainty has been seized upon by many as a justification for inaction. Nothing could be more foolish than doing nothing. If global warming is a real phenomenon, it will not be reversible. If it is real it may also give rise to some extreme and locally catastrophic outcomes. Unless we are risk lovers — and few of us behave as if we are — this combination of uncertainty, irreversibility and potentially large-scale effects points in only one direction — caution. On that basis alone we have adequate rationale for policies to reduce greenhouse gas emissions and to invest in biomass enhancement to sequester carbon.

As it happens, the rationale is stronger than this. For the policies needed to reduce the rate of global warming themselves have numerous side-benefits. If we slow the rate of tropical deforestation in order to slow the rate of carbon emissions from forest burning, we also help conserve the biological diversity in the tropical forests. We get at least a 'double benefit'. If we reduce carbon dioxide emissions through energy conservation, switching to low carbon fuels or to renewable energy sources, we also reduce other pollutants — sulphur and nitrogen oxides for example. If we tackle carbon dioxide emissions in the transport sector the policies that are likely to be used also reduce congestion, noise, accidents and local pollution. Finally, while politicians bemoan the cost of global warming control (for reasons of genuine concern or pure political strategy) it is, in reality, not expensive at least to begin aggressive measures.

The balance of costs and benefits therefore favours policies of firm control over greenhouse gases. But convincing people without resorting to scare-mongering and fanciful predictions requires a rigorous and painstaking analysis of what we know about the likely effects of global warming. Building on precedents set by William Cline and William Nordhaus in the USA, Sam Fankhauser has produced the most careful analysis yet of the likely economic damages arising from global warming. He is careful to provide regional estimates of damage as well as global costs. The stake of the developing countries in global warming control turns out to be larger than they themselves — small island states and deltaic countries apart — might have supposed. This suggests that the Framework Convention on Climate

Change, agreed at Rio de Janeiro in 1992, will have to adopt tougher and more comprehensive targets for its next phase beyond the year 2000. There is growing evidence that politicians will be reluctant to agree to stricter targets. Some firmer scientific conclusions by then may persuade them otherwise.

Sam Fankhauser shows that measurable damage costs of doubling CO_2 concentrations are around 1 per cent to 2 per cent of gross 'world' product (GWP). This benchmark figure has to be compared to the cost of reducing emissions. These may amount to 1 per cent to 3 per cent of GWP, for a 50 per cent cut by 2025–50, if 'top down' costing models are used, but less than this if 'bottom up' models are used. The ranges of costs and benefits overlap, but they are not strictly comparable. A 50 per cent emissions cut will not necessarily result in a benefit of 1 per cent to 2 per cent of GWP, since it may not succeed in fully avoiding concentration doubling. On the other hand, as noted, the comparison omits the ancillary benefits of climate change action, and those are potentially very large. The balance therefore favours those who call for tougher action.

In the space of only a few years, Sam Fankhauser's work has gained worldwide recognition. This book is a summary of that work and CSERGE is proud to have been the environment in which it was developed.

David Pearce
Director, Centre for Social and Economic Research on the Global Environment, University College London and University of East Anglia.

Preface

This book is about economics and climate change. In particular, it emphasises an aspect which has gained little attention in the debate so far: the economic valuation of climate change impacts.

Greenhouse economics is a quickly evolving field. Rapid progress makes it certain that most contributions have only a very short lifetime, and this volume will undoubtedly be no exception. It is symptomatic in this context that most of the literature cited in this book is less than three or four years old. While this is perhaps bad news for authors, whose work is outdated almost by the time of printing, it is good news for the issue at hand. An improved understanding of the climatological and socio-economic consequences of global warming and global warming policies is urgently needed. Greenhouse damage research, in particular, is a field where very little is known and where much more research input is warranted. The merit of this volume lies thus not so much in providing final answers – it does not – but in making a contribution to work in progress, and in outlining directions for future research.

Most of the material presented in this book is based on my PhD dissertation at University College London on *Greenhouse Economics and the Costs of Global Warming*. During the course of my studies at UCL I have profited from the help of many people and institutions. For one year my work was financed by a stipend of the Schweizerischer Nationalfonds für Forschung und Wissenschaft, and their support is gratefully acknowledged. Subsequently, I joined the Centre for Social and Economic Research on the Global Environment (CSERGE) at UCL. My work has greatly benefited from the stimulating environment at CSERGE, and I am grateful to my colleagues at both UCL and the University of East Anglia, where the second branch of the Centre is located. I am above all indebted to Professor David Pearce, my supervisor and the Centre's director at the London end. The financial support extended by the UK Economic and Social Research Council (ESRC) to the Centre is also gratefully acknowledged.

As the work evolved, numerous people have commented on different aspects and draft versions of the book, and I am grateful to Neil Adger, Rosemary Clarke, William Cline, Emily Fripp, Robert Kay, Joaquim Oliveira Martins, William Nordhaus, Richard Richels, Jens Rosebrock, Joel B Smith and R Kerry Turner for their comments. I have particularly profited from the thoughtful comments by Alistair and David Ulph, joint directors of the CSERGE global warming programme, and from the discussions with my two examiners, Scott Barrett and John Proops.

I am also extremely grateful for the encouragement I got from my friends at UCL and elsewhere: Pat Fairbrother, Viju James, Esra Karadeniz, Cuneyt Karadeniz, Guido Kürsteiner, Ana Rico Gomez, Marcel Rindisbacher, Archana Srivastova, Anca Toma, Jennifer Williams and particularly Snorre Kverndokk, who was a visitor at UCL in 1990/91, and with whom I have collaborated since. The greatest thanks of all I owe to my parents, who throughout my studies have encouraged and supported me in the most wonderful way. I dedicate this book to them.

London, July 1994

PART I

Introduction and Overview

◆

Chapter 1

Economics and Climate Change

THE EMERGENCE OF THE GREENHOUSE PROBLEM

Global warming history

Within only a few years global warming has transformed from a scientific speculation to an environmental threat of world-wide concern. The global awareness is underlined by the fact that over 160 parties, representing virtually every nation in the world, have signed the UN Framework Convention on Climate Change which was initiated at the Earth Summit in June 1992, in Rio de Janeiro (see 'global warming politics' below).

While the general public has only recently become aware of the global warming problem, the scientific greenhouse debate covers a span of more than 150 years. In 1824 French scientist Jean Baptiste Fourier first described the *natural* greenhouse effect, drawing a parallel between the action of the atmosphere with the effect of glass covering a container. As is now well known, the natural greenhouse effect described by Fourier forms an important part of the earth's energy balance system and is essential for life on earth: without it the average surface temperature on the planet would be a mere $-18°C$, rather than the $15°C$ observed today – too low for any sort of life. The possibility of an *enhanced* or *man-made* greenhouse effect was introduced some 70 years later, in three articles by the Swedish scientist Svante Arrhenius published around the turn of the century (Arrhenius, 1896; 1903; 1908). Arrhenius hypothesized that the increased burning of coal, which had paralleled the process of industrialization, may lead to an increase in atmospheric CO_2 concentration and warm the earth. He believed, however, that most of the anthropogenic CO_2 would be absorbed by the oceans, and that, in any case, warming would be desirable. Atmospheric CO_2 concentration at that time was still below 300 ppm (parts per million by volume).

Throughout the following decades the greenhouse effect continued to be a topic of moderate scientific interest, but of little general concern. In 1965 President Johnson's Science Advisory Committee included a chapter on atmospheric CO_2 in a report on environmental problems. It was the first official US government document to mention the issue. The concentration of atmospheric CO_2 was now continuously monitored at the Mauna Loa

Observatory in Hawaii and at the South Pole. By 1965 it had risen to about 320 ppm. Yet the concern in the early 1970s was less about global warming than about global cooling, induced by industrial and agricultural aerosols. The concern about global warming was again revived at the First World Climate Conference in Geneva in 1979. The conference called for increased efforts towards a better understanding of climate change, but, given the low consensus at the time, was still cautious about the issue.

The turning point was an international conference in Villach, Austria, in 1985. Sponsored by the United Nations Environment Programme (UNEP), the World Meteorological Organisation (WMO) and the International Council of Scientific Union (ICSU), the conference achieved a consensus among scientists about the magnitude of the problem, noting that

> [a]s a result of the increasing concentrations of greenhouse gases ... in the first half of the next century a rise of global mean temperature could occur which is greater than any in man's history' (conference statement, quoted in Jäger, 1992).

Global warming was established as a problem of international concern, at least in the scientific community. Atmospheric CO_2 concentration had reached 345 ppm.

Other workshops followed. The sponsors of the Villach conference had established an Advisory Group on Greenhouse Gases (AGGG), which organised two follow-up workshops in Bellagio (Italy) and again in Villach in 1987. The World Conference on the Changing Atmosphere held in Toronto in 1988 called for a reduction in global emissions of at least 50 per cent to stabilize the atmospheric concentration of CO_2. As an initial goal it proposed an emissions reduction of 20 per cent compared to 1988 levels, to be achieved by 2005. The conference enjoyed high media attention, maybe not least because at the same time the United States experienced one of the worst droughts ever. Despite the non-official character of the conference and the arbitrariness of the target, the 'Toronto goal' of a 20 per cent emission cut was soon established as an abatement benchmark which has been used in much of the discussion since. Another well-known target, calling for a warming limit of 0.1°C per decade and 1–2°C in total, was put forward by one of the AGGG working groups. Not surprisingly, the political community found it rather difficult to endorse such targets, mainly because they were usually issued without much consideration being given to the costs involved in achieving them. Governmental meetings, like the Ministerial Conference on Atmospheric Pollution and Climate Change held in Noordwijk (Netherlands) in 1989, invariably failed to agree on emission targets. The discrepancy between science and policy became manifest at the Second World Climate Conference in 1990, held again in Geneva, where scientists were reiterating their call for strong and immediate action, while the ministerial conference held immediately afterwards produced very few results. Meanwhile atmospheric CO_2 concentration had reached 353 ppm.

Box 1.1
Global warming chronology

1824   Jean Baptiste Fourier first describes the greenhouse effect, comparing the action of the atmosphere to that of glass covering a container.

1850–70   Industrial revolution intensifies. Atmospheric CO_2 concentration is at 285 ppm.

1896, 1903,   In three articles the Swedish scientist Svante Arrhenius hypothesises
1908   that burning coal will increase CO_2 and warm the earth. He suggests that warming may be desirable.

1958   Continuous monitoring of CO_2 concentrations in the atmosphere is started at the Mauna Loa Observatory on Hawaii and at the South Pole.

1965   The US President's Science Advisory Committee includes a chapter on atmospheric CO_2 in its report on environmental problems. It is the first official US government report on the problem.

Early 1970s   Widespread concern develops over potential global climatic cooling induced by industrial and agricultural aerosols.

1979   The first World Climate Conference is held in Geneva. Concern is revived, although the conference statement is cautious about the issue of climate change.

1985   The international conference in Villach (Austria) clearly establishes greenhouse warming as an international concern. An Advisory Group on Greenhouse Gases (AGGG) is established.

1987   Under the auspices of AGGG two workshops are held, in Villach and Bellagio (Italy), which further increase the world-wide attention on greenhouse warming.

1988   The World Conference on the Changing Atmosphere held in Toronto calls for a reduction in CO_2 emissions by 20% of 1988 levels by the year 2005. In November UNEP and WMO establish the Intergovernmental Panel on Climate Change (IPCC).

1990   Second World Climate Conference in Geneva on the results of IPCC. Strong call for action in response to climate change. The scientific/technical conference is followed by a ministerial conference.

1991   Negotiations start for a Framework Convention on Climate Change.

1992   UN Conference on Environment and Development (UNCED) in Rio de Janeiro. The Climate Convention is signed by over 160 nations. No legally binding targets, but expression of the aim to reduce emissions to 1990 levels.

1994   Three months after the fiftieth ratification, the Framework Convention comes into force.

Source: Handel and Risbey (1992); Jäger (1992)

Global warming politics

With the Second World Climate Conference global warming entered the political sphere. The conference was again a big media event. Politicians were keen on indicating their awareness of the issue. In her address, the UK's Prime Minister Margaret Thatcher reminded the conference about man's 'duty to nature', and emphasized the need for 'a successful framework convention on climate change'. Similar phrases could be heard from most other political leaders present in Geneva (see Jäger and Ferguson, 1991).

In February 1991 international negotiations began for a UN Framework Convention on Climate Change. Over the course of the negotiations, stringent proposals like the Toronto target were soon replaced by more modest objectives. The convention, which was finally signed at the 'Earth Summit' – the UN Conference on Environment and Development – in Rio de Janeiro in June 1992, states as the final aim to stabilize greenhouse gas concentrations 'at a level that would prevent dangerous anthropogenic interference with the climate system' and this 'within a time frame sufficient to allow ecosystems to adapt naturally to climate change' (Article 2). It does not, however, contain a binding commitment of any sort and merely urges the so called Annex 1 countries[1] to reduce emissions to 1990 levels by the year 2000 – the more modest Rio target which has replaced the Toronto call of 1988. The Convention also contains provisions for financial assistance for the implementation of abatement or protection measures in developing countries.[2]

The Climate Convention has been signed by over 160 countries so far. The fiftieth ratification was received in December 1993, and three months later, in March 1994, the Convention came into force. The list of the first 50 ratifying nations includes about half of the Organization for Economic Co-operation and Development (OECD) member states, important developing countries like China and India as well as many of the small island states, who will be most vulnerable to climate change (see Box 1.2). The European Union is not among the first 50 nations. It ratified the convention as 52nd party just before Christmas 1993. Most OECD countries have now committed themselves in some form to the Rio target of CO_2 emissions stabilization at 1990 levels, although not all these commitments appear to be firm (see Table 1.1).

Global warming research

Meanwhile, global warming research is strongly in the hands of the Intergovernmental Panel on Climate Change (IPCC). The IPCC was initiated by

---

1 Annex 1 countries include the economies in transition and the industrialized countries of the West.
2 For an excellent summary and assessment of the Rio outcome, also including the other agreements signed, see Grubb et al (1993).

Box 1.2
The first 50 ratifying nations of the Climate Convention

1. Mauritius	26. Iceland
2. Seychelles	27. Uzbekistan*
3 Marshall Islands	28. Dominica
4. United States of America	29. Sweden
5. Zimbabwe	30. Norway
6. Maldives	31. Tunisia
7. Monaco	32. Burkina Faso
8. Canada	33. Uganda
9. Australia	34. New Zealand
10. China	35. Mongolia
11. Saint Kitts and Nevis	36. Czech Republic
12. Antigua and Barbuda	37. Tuvalu
13. Ecuador	38. India
14. Fiji	39. Nauru
15. Mexico	40. Jordan
16. Papua New Guinea	41. Micronesia
17. Vanuatu	42. Sudan
18. Cook Islands	43. Sri Lanka
19. Guinea	44. United Kingdom
20. Armenia	45. Germany
21. Japan	46. Switzerland
22. Zambia	47. Republic of Korea
23. Peru	48. Netherlands
24. Algeria	49. Denmark
25. Saint Lucia	50. Portugal

* Accession

Source: United Nations, Intergovernmental Negotiating Committee.

---

the United Nations Environmental Programme (UNEP) and the World Meteorological Organization (WMO) in 1988 with the triple task of: (a) furthering scientific understanding about global warming (Working Group 1), (b) assessing its impacts (Working Group 2); and (c) formulating possible response strategies (Working Group 3). By the time of the Second World Climate Conference, when it presented its first report, the IPCC had become the authoritative body on global warming. Several hundred scientists had been involved in producing the reports of the three Working Groups, either as authors or as reviewers. The IPCC succeeded not only in achieving a scientific consensus but also in conveying the issue to a hitherto un-informed or ill-informed public (Grubb et al, 1993). Its work continued after the conference and a supplementary report was issued in 1992. A second round assessment is being carried out throughout 1994, with the aim of producing an updated assessment by early 1995, in time for the first

Table 1.1 Greenhouse gas emission targets of OECD countries

Country	Gases Included	Sectors Included[1]	Action	Base Year	Commitment Year	Conditions/Comments
Australia	NMP GHGs	All	Stabilization 20% reduction	1988 1988	2000 2005	Interim planning target conditional on 'no regrets' measures and similar action by other GHG-producing countries.
Austria	CO_2	All	20% reduction	1988	2005	
Belgium	CO_2	All	5% reduction	1990	2000	
Canada	CO_2 and other NMP GHGs	All	Stabilization	1990	2000	Target is to stabilize the aggregate global warming potential of the gases included.
Denmark	CO_2	All	20% reduction	1988	2005	Target for the transport sector alone is to stabilize CO_2 emissions by 2005 and to achieve a 25% reduction by 2030.
Finland	CO_2	Energy	Stabilization		End 1990s	Target is to stabilize emissions at below two metric tons of carbon per capita per year by 2000.
France	CO_2	Energy	Per capita stabilization		2000	
Germany	CO_2	Energy	25%–30% reduction	1987	2005	Target is 25%–30% reduction in CO_2 emissions. There is no official target for other NMP GHGs, but the government is striving to reduce the overall warming potential of total GHG emissions by 50% from 1987 levels by 2005.
Greece	CO_2		(EU Agreement)[2]			
Iceland	All GHGs	All	Stabilization	1990	2000	
Ireland	CO_2	All	Limitation to 20% growth	1990	2000	Net CO_2 emissions are expected to grow by 11% between 1990 and 2000.
Italy	CO_2	All	Stabilization	1990	2000	
Japan	CO_2	All	Per capita stabilization	1990	2000	The government has also stated that efforts should be made to stabilize total CO_2 emissions beyond 2000 at about the same level as in 1990.
Luxembourg	CO_2 "	All "	Stabilization 20% reduction	1990 1990	2000 2005	
The Netherlands	CO_2 " All other GHGs	All " All	Stabilization 3–5% reduction 20%–25% reduction	1989–1990 1989–1990 1989–1990	1994–1995 2000 2000	 3% reduction; 5% reduction depending on international developments and opportunities. Gas-by-gas strategy.

Country	Gases Included	Sectors Included[1]	Action	Base Year	Commitment Year	Conditions/Comments
New Zealand	CO_2	All	Stabilization	1990	2000	Primary objective; the ultimate objective is to reduce CO_2 emissions by 20% from 1990 levels by 2000 conditional on 'no-regrets' measures.
Norway	CO_2	All	Stabilization	1989	2000	Preliminary target
Portugal	CO_2		(EU Agreement)[2]			
Spain	CO_2	Energy	Limitation to 25% growth	1990	2000	
Sweden	CO_2	Energy	Stabilization	1990	2000	
	"	"	Reduction	1990	After 2000	
Switzerland	CO_2	Energy	Stabilization	1990	2000	
	"	"	Reduction	1990	After 2000	
Turkey			(No target has been set)			
United Kingdom	CO_2, methane, other major GHGs	All	Stabilization	1990	2000	Gas-by-gas approach; specific targets are set for different gases.
United States	All GHGs	All	Stabilization	1990	2000	Target is to stabilize the aggregate global warming potential of all GHGs.
EU	CO_2	All	Stabilization	1990	2000	Target is for Union as a whole.

Notes: NMP = Non-Montreal Protocol. Refers to greenhouse gases other than those covered under the 1987 'Montreal Protocol on Substances that Deplete the Ozone Layer' and its subsequent amendments, ie, greenhouse gases other than CFCs, HCFC, halons, carbon tetrachloride, and methyl chloroform. GHGs = Greenhouse gases.
1) Targets for the energy sector include emissions from energy use in transport, unless otherwise stated.
2) 'EU Agreement' means the country in question falls under the EU-wide target stated at the end of the table but has not yet developed its own target.

Source: International Energy Agency, (1994)

meeting of the Conference of Parties. In this second assessment socio-economic aspects are given a much stronger weight: the reorganized Working Group 3 is now predominantly devoted to socio-economic considerations.

The questions of how economics can contribute to the global warming discussion and what the state of the art of greenhouse economics is, will be the main topic of this book. The remainder of this introductory chapter outlines the structure of the book and sums up the main conclusions.

OUTLINE AND SUMMARY

There is a relatively large body of literature on many economic aspects of global warming, in particular on greenhouse gas abatement costs. The question of greenhouse damage valuation, on the other hand, has gained little attention so far. Part II is devoted to a further analysis of this issue. Chapter 2 surveys our current knowledge about the costs of global warming damage, and discusses the main shortcomings of existing studies. The following chapters then contain extensions to these studies, which try to meet at least some of the criticisms raised.

Chapter 3 deals with the case of atmospheric CO_2 concentration doubling, extending existing US damage assessments to other geopolitical regions and the world as a whole. The estimates confirm the often used damage cost benchmark of 1 per cent to 2 per cent of GNP, at least for developed countries. Estimates for developing countries are about 60 per cent higher, supporting the view that the poorest countries will suffer most, even if adequate adaption measures are taken. Regional differences may, however, be considerable, as will be exemplified by the estimates for the former Soviet Union and China.

Chapter 4 transforms these equilibrium, *total damage* figures into *marginal costs*, that is into a benchmark number of damage caused per tonne of emission. The calculations are based on a stochastic greenhouse damage model in which all key parameters are random. This, on the one hand, allows a closer representation of current scientific understanding, on the other hand it will enable us to calculate a damage probability distribution, and thus to account explicitly for the uncertain nature of the global warming phenomenon. As a benchmark, the model calculates that CO_2 emissions impose social costs in the order of 20 \$/tC for emissions between 1991 and 2000, a value which is estimated to rise over time to reach about 28 \$/tC by 2021–30. Similar figures for CH_4 and N_2O are also provided.

The magnitude of greenhouse damage does not only depend on the amount of warming encountered, but also on the way people adjust to it. Chapter 5 illustrates the importance of adaptation measures with the example of sea level rise. The chapter introduces a rule of thumb to approximate the optimal level of coastal protection, based on the criterion of economic efficiency. As is intuitively apparent, the optimal protection level under an efficiency criterion will depend on the relative magnitude of land-loss costs in the absence of protection, on the one hand, and the costs of full protection, on the other. The total costs of sea level rise then consist of two parts: the costs of protection, on the one hand; and the costs of the remaining, unmitigated damage, on the other. This framework is used to estimate sea level rise damage cost functions for the countries of the OECD.

Part III links the question of global warming damage estimation with the two broader questions of optimal policy design and international coopera-tion. The first two chapters of Part III recapitulate the state of the art on the abatement cost side. Chapter 6 deals with energy and abatement cost models, and Chapter 7 with policy instruments. Energy modelling has probably been the most prominent aspect of greenhouse economics so far, or at least the one economists got involved in first. Consequently, there is now an abundance of different models and model results. An emerging rule of thumb appears to be that a 50 per cent reduction in baseline CO_2 emissions, achieved by 2025 to 2050, would cost about 1 per cent to 3 per cent of GNP, at least according to one type of models, the so called *top down* models. The other main type of models, *bottom up* models, consistently predict lower costs, though. Bridging the gap between the two schools is therefore one of the main tasks for energy modellers in the future. Net abatement costs will also be lower, once two positive spillover effects are taken into account: the side-effect of lower air pollution levels (the so-called *secondary benefits*) and the efficiency gain from replacing existing taxes with a carbon tax (the so-called *double dividend*).

Because of the long-term character of global warming, the choice of the discount rate plays a crucial role in the climate change debate. The most significant global warming impacts will probably not be felt before the middle of the next century. If greenhouse damages in, say, the year 2100 are then discounted at 6 per cent rather than 3 per cent, their present value is reduced by about a factor 20. Not surprisingly, the discussion about the correct discount rate is therefore fierce. The main arguments are summar-ized in Chapter 8. There are two main schools of thought. The first would set the rate of discount in accordance to observed economic behaviour. That is, the discount rate is fixed at the level which allows economic models to replicate past savings and interest rate trends. The second school takes a more prescriptive, rather than descriptive, stance and asks what the ethically correct discount rate should be. It usually argues for a much lower rate than the descriptive school, on the basis of moral principles such as inter-generational equity.

Chapter 9 puts the previous considerations on discounting, damage and abatement costs together and asks about the optimal policy response to global warming. While in general advocating a cost-benefit approach, the chapter does not hide the potential shortcomings of this method. Problems particularly occur with respect to ethical considerations and the treatment of climate uncertainty. We also call for a more comprehensive view on global warming policy which considers all aspects relevant to the problem, including, for example, not just CO_2 but all greenhouse gases, and accounting for side-effects on other externalities (secondary benefits and double dividend, as noted above).

Chapter 10 raises issues of international cooperation. The question of how much and what type of global warming action to take represents only one aspect of the debate. Deciding on who should carry the abatement burden is a second key issue. Chapter 10 inquires who the likely gainers and losers would be under different types of international agreements. It also discusses the usefulness of unilateral action in the face of what is called *carbon leakage*. The term refers to a situation in which the direct effect of reduced emissions in abating countries is partly offset by increased emissions from non-signatories. To what degree leakage would occur in the real world is still an open question, and model predictions vary considerably.

Conclusions about the state of the art of global warming economics are drawn in Chapter 11.

PART II

The Economic Costs of Global Warming

——— ◆ ———

Chapter 2
Overview: Global Warming Damage

INTRODUCTION

Since the early days of the greenhouse debate scientists have been interested in the impacts of global warming. Working Group 2 of the Intergovernmental Panel on Climate Change (IPCC) has been exclusively devoted to this topic (IPCC, 1990b). In the United States the Environmental Protection Agency has initiated a comprehensive study on the impacts of climate change for the country, the results of which are described in the extensive report by Smith and Tirpak (1989). The work of these two groups has been complemented by a multitude of additional studies including, for example, Parry et al (1988) on agriculture, Waggoner (1990) on water, Peters and Lovejoy (1992) on biological diversity, and the World Health Organisation (1990) on health effects, to name only a few. This large interest is not really surprising, after all it is the fear of potentially devastating consequences which has put global warming onto the agenda. Nevertheless, attempts at a monetary quantification of these impacts – despite being a classic application of environmental economics – have started to emerge only recently. Partly this can be attributed to the fact that proper economic valuation requires a reasonably accurate knowledge of impacts in physical terms, something which even now is available only to a limited extent.[1] It should also be clear that the valuation of such damage aspects as human hardship will push economic valuation techniques to their limit, and quite possibly beyond. Yet many aspects *are* quantifiable in monetary terms, and damage valuation is thus probably the one area in greenhouse research where most catching up is required.

In classifying greenhouse damage models we can distinguish between two approaches. The first is a partial equilibrium approach. Each damage aspect is analysed separately and total greenhouse damage is simply the sum of the individual categories. This method is often called the *enumerative approach* to greenhouse damage estimation (Cline, 1994). The second method analyses greenhouse damage in a more *integrated*, general equilibrium set up.

---

1 Models which provide a reasonably detailed assessment of greenhouse damage even at a regional level include IMAGE and ESCAPE, see Rotmans et al (1990) and Climate Research Unit (1992).

That is, climate change is imposed on a system of interacting markets. Initial impacts on one sector may then also have higher order effects and spill over to other sectors of the economy.

We will deal with each of these methods in turn. In doing so, the main focus is on comprehensive assessments which consider greenhouse damage as a whole, and less so on work concentrating on only a particular aspect of the problem.

GREENHOUSE IMPACTS

Global warming will have a variety of effects (see Figure 2.1). They can be classified as either market related – ie effects which will become manifest in the national accounts – or non-market related – ie impacts affecting 'intangibles', such as ecosystems or human amenity.

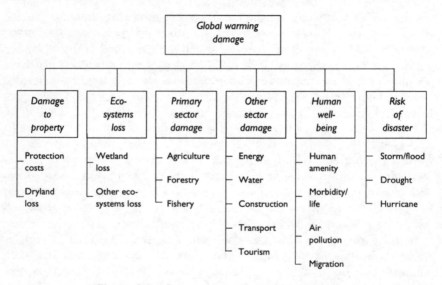

Figure 2.1   Overview on global warming impacts

The Intergovernmental Panel on Climate Change (IPCC) foresees a shift in the current agricultural production pattern away from current production areas to more northern latitudes. Together with changes in soil water availability, the increased occurrence of climatic extremes and crop diseases, this may lead to an overall reduction in agricultural yields, and could result in serious regional or year-to-year food shortages (Parry, 1993). The IPCC further predicts that the increased stress on ecosystems may lead to the extinction of species unable or too slow to adapt.

The rise in sea levels connected with a warmer climate will threaten low-lying coastal areas. Sea level rise will particularly affect densely populated coastlines and small island states. Extreme events like floods and droughts may occur more frequently. Health experts expect a rise in climate-related diseases such as heat strokes and a spread of vector-borne diseases like malaria into areas so far unaffected. Others have warned about the consequences of increased water shortages in summer. Climate-dependent economic activities like construction, transport and tourism will be affected in an unknown way. Finally, global warming may trigger a stream of climate refugees away from the worst-affected regions and coasts (IPCC, 1990b).

The scientific research on global warming impacts has almost entirely focused on the case of CO_2 concentration doubling, that is on the impacts of an atmospheric CO_2 concentration of twice the preindustrial level ($2xCO_2$). $2xCO_2$ is a completely arbitrary benchmark, chosen solely for analytical convenience. It is neither an optimal point nor a steady state, and warming will continue, and in fact aggravate beyond $2xCO_2$. This point has repeatedly been made in the literature, most prominently by Cline (1992a). Nevertheless, as a consequence of the scientific focus, studies on the economic costs of global warming have also tended to concentrate on $2xCO_2$. By far the best studied aspects are the impacts on agriculture and the costs of sea level rise.[2] Some further studies exist on, for example, forestry, but hardly any attempts have been made to assess other aspects, particularly the damage to non-market sectors. Several authors have nevertheless tried to provide a first order assessment of total global warming damage, including non-market aspects.[3]

ENUMERATIVE STUDIES

The pioneering paper trying to provide a first order assessment of greenhouse damage in economic terms was the well-known study by Nordhaus (1991b, c). Concentrating mainly on the costs of agriculture and sea level rise, he estimated an overall damage of global warming in the order of a quarter per cent of GNP. To allow for the many non-market impacts neglected in the study, this value is raised to 1 per cent, with a range of error of 0.25 per cent to 2 per cent. The figures are based on US data, but Nordhaus claims that they may hold world-wide. Improvements on Nordhaus' back-of-the-envelope estimate have been provided by Cline (1992a) and Titus (1992), two studies again focusing on the US. Tol (1993) distinguishes between several world regions. These studies all perform the same hypothetical exercise, and ask what the damage would be if today's economy were exposed to $2xCO_2$. An overview of results for the United States is provided in Table 2.1. The individual damage aspects will be discussed in more detail in Chapter 3.

---

2 Agricultural studies include, eg, Adams et al (1990), Kane et al (1992), and Rosenzweig and Parry (1994). For sea level rise cost assessments see, eg, Fankhauser (1994b), Titus et al (1991) and Rijsberman (1991).

3 Cline (1992a), Nordhaus (1991b, c), Titus (1992) and Tol (1993).

Table 2.1   Economic damages from 2xCO$_2$ for the US (present scale economy, bn$)

	Cline (1992a)[a] (2.5°C)	Titus (1992)[a] (4°C)	Tol (1993)[b] (3°C)	Nordhaus (1991b, c)[a] (3°C)
sea level rise	6.1	5.0	8.5	10.7
agriculture	15.2	1.0	10.0	1.0
forest loss	2.9	38.0	–	small
fishery	–	–	–	small
energy	9.0	7.1	–	1.0
water supply	6.1	9.9	–	
other sectors	1.5[c]	–	–	
ecosystems loss	3.5	–	5.0	
human amenity	–	–	12.0	
life/morbidity	5.0	8.2	37.4	d
migration	0.4	–	1.0	
air pollution	3.0	23.7	–	
water pollution	–	28.4	–	
natural hazards	0.7	–	0.3	
Total (bn$)	53.4	121.3	74.2	48.6
(% GNP, 1988)	(1.1)	(2.5)	(1.5)	(1.0)

a transformed to 1988 *values* based on per cent GNP estimates
b USA and Canada
c tourism, urban infrastructure
d not assessed categories, estimated at 0.75 per cent of GNP

Sources: as indicated.

Despite differences in individual damage categories – and not all papers deal with all aspects – the studies display a surprisingly high degree of consistency in the overall result. The emerging picture from enumerative studies seems to be that, to the extent that it can be monetized, *2xCO$_2$ damage in developed countries can be expected to lie in the order of 1 per cent to 2 per cent of GNP*, for a 2.5–3°C warming. In as far as the results are comparable, this benchmark figure is roughly confirmed by a poll of impact experts conducted by Nordhaus (1993d, 1994), although the Nordhaus study exhibits a much wider range of opinions. Asked about the likely impacts of a 3°C warming by the year 2090, experts predicted a damage in the order of 0.7 per cent to 8.2 per cent of GNP (90 per cent confidence interval). The median value was 1.9 per cent.

The figures of Table 2.1 show considerable differences with respect to individual damage categories. Agricultural damage according to Cline

(1992a), for example, is more than an order of magnitude higher than the respective figures of Nordhaus and Titus. The Titus study, on the other hand, reports pollution and forestry damages significantly higher than the other assessments.

There are broadly two reasons for such differences. Firstly, there are discrepancies due to different assumptions about the quantitative impacts. These emerge mainly as a consequence of scientific uncertainty on the exact physical impacts of global warming. Agricultural damage is probably the best example for this. Secondly there are cases where the authors agree on the quantitative impacts but value them differently. The forestry estimates are a prime example here. Another one is mortality, where surveys report a range of $0.2 m to $16 m for the value of a statistical life (Pearce et al, 1991; Viscusi, 1993). Cline's assessment in Table 2.1 is based on a rather low value of $0.6 m, while Tol (1993) uses values of about $3 m (for developed countries). The point to note is that ignorance with respect to valuation may cause a similar range of error as scientific uncertainty, and that the quality of estimates can be ameliorated by improving either side. This conclusion is confirmed by Peck and Teisberg (1993b), who have calculated the value of improved information, and found similar figures for a better knowledge about the costs of warming as for an improved scientific understanding (see Table 9.1 below).

INTEGRATED DAMAGE ASSESSMENTS

In the enumerative approach, total damage is the sum of individual damage categories, ie the estimates are based on a partial equilibrium approach. *Inter alia* this means that they neglect the higher order effects which, for example, a change in agricultural yields will induce on the food, tobacco or textile industry. Integrated damage studies, or general equilibrium assessments, correct for this shortcoming. However, while more comprehensive with respect to the treatment of market based damages, they usually ignore non-market impacts.

On an abstract level the difference between a partial and general equilibrium approach has been analysed by Kokoski and Smith (1987), without however achieving any general results other than that the error from using the partial equilibrium method may be 'quite large' (p 340).

On a more applied level general equilibrium effects have been incorporated in the so-called MINK study, probably the most comprehensive regional damage estimation project undertaken so far.[4] Concentrating on agriculture, forestry, water and energy, the study analyses the impacts of climate change on the four US states Missouri, Iowa, Nebraska, and Kansas (hence the acronym MINK). First order damage effects, eg on agriculture,

---

4 For a summary see Rosenberg and Crosson (1991). A more detailed description of the project can be found in the Special Issue of *Climatic Change* 24(1/2), 1993.

are fed back into the economy through an input-output module. The study has other interesting features. Rather than working with $2xCO_2$ data, the study uses the weather pattern of the 1930s as a climate analogue, thus allowing for more temporal and spatial variability. Estimates are provided for both today's economy and that expected for 2030. Particular care is further taken to the issue of adaptation. The study concludes that global warming would probably not reduce regional income by more than 1 per cent (for roughly 1°C warming). The impacts on agriculture and water would be most strongly felt.

In a similar attempt, Scheraga et al (1993) have used the Jorgenson and Wilcoxen (1990) general equilibrium model to estimate the macroeconomic effects of climate change for the US. The study is limited to three types of impacts: a change in agricultural yields, a change in energy demand, and an increase in the sea level. Estimates are based on a rather pessimistic $2xCO_2$ scenario of 5.1°C warming, assumed to occur gradually between now and the year 2060. The study illustrates how the structure of an economy may adjust to climate change by moving away from agriculture and consumption-related activities towards investment and capital-related industries. The increase in public expenditures on sea level rise protection, that is an increase in government purchases from the construction sector, leads to a redirection of spending away from consumption towards investment. At the same time higher prices for agriculture-based products such as food and tobacco lead to a fall in the demand for these goods. Overall, GNP in the year 2060 is 0.8 per cent lower than in the base case without warming. In comparison, agriculture, sea level rise and energy account for about 0.2 per cent to 0.6 per cent of GNP in the enumerative studies, although assuming only about half as much warming (see Table 2.1).

SHORTCOMINGS AND EXTENSIONS

Valuation

Comprehensive damage assessments have been fiercely criticised by many authors.[5] Probably the main objection concerns valuation, in particular the validity of economic valuation techniques. Grubb (1993) for example criticised $2xCO_2$ damage estimates as being based on a 'largely subjective valuation of non-market impacts' (p 153). However, while some of the existing figures are indeed based on rather weak assumptions, it would be wrong to take this as an indication of the ineptitude of economic valuation techniques in general. Economic valuation of non-market goods is controversial. Nevertheless, the problem of greenhouse damage estimates is currently perhaps not so much the accuracy of valuation methods as such,

---

5 Eg Grubb (1993), Fisher and Hannemann (1993), Ayres and Walter (1991), Daily et al (1991) and Morgenstern (1991).

but the fact that they have not yet been applied to the problem to a sufficient degree. This is not to say that a full and complete valuation of all greenhouse impacts will ever be possible. Given the size of the problem and the uncertainties involved, it will probably not be, at least not within reasonable time. However, existing estimates are clearly far from perfect and the policy debate would gain from their improvement. Further and more detailed valuation studies are thus warranted. What is particularly needed is a broadening of the scope from the emphasis on agriculture and sea level rise to the inclusion of other damage aspects such as ecosystems loss, climate amenity, health and morbidity.

Global vs regional assessment

Better estimates are also needed with respect to the damage costs to developing countries, where next to nothing is known about the willingness to pay for non-market goods like wetlands, for example, or the costs of mortality. Many authors have questioned the Nordhaus claim that his US estimates can be extended to the world as a whole (eg Grubb, 1993). Indeed one would intuitively expect impacts to be more severe in developing countries, where the dependence on climate-related sectors is considerably higher than in the first world. This point will be further analysed in Chapter 3, and the results will on the whole confirm this intuition. Special regions like the former Soviet Union apart, damage (relative to GNP) to developing countries may be about 60 per cent higher than the OECD average. Tol (1993) even obtains damage levels up to 4 times higher than the OECD average for some developing regions. The situation for LDCs could be further aggravated by a failure to implement the cost efficient adaption responses (eg coastal protection), something which is quite likely to happen if the necessary funds are not made available. It is all the more important therefore to study the impacts on developing countries more carefully.

Adaptation

Most authors agree, in principle, that damage estimates should be based on the assumption of a cost efficient response. That is, damage should be kept at a minimum through appropriate adaptation measures. This may, for example, include the erection of sea defences, the development of heat resistant crops, a change in agricultural and forest management, the construction of water storage and irrigation systems, the adaptation of houses, and the like. Schelling (1992) has even argued that for developing countries the best adaptation strategy may simply be economic development. As we will show in Chapter 5, *the costs of global warming then consist of two parts: the cost of adaptation; and the costs of remaining, unmitigated damages.* Unfortunately, while undisputed as a general idea, data limitations do not always allow the implementation of optimal adaptation strategies when it comes to actual damage calculations. For simplicity it is then often assumed that no adjustments are taken at all, or that current service levels (eg air quality standards) are maintained at all costs.

The failure to capture managerial responses in much of the agricultural damage literature may serve as a prime example here. The assumption implicit in many earlier studies is that farmers will continue to plant the same crops, even though climatic conditions have changed – the 'dumb farmer' hypothesis, as this assumption is often termed. Recent studies have corrected for this shortcoming and incorporate managerial responses into agricultural models.[6] Their results seem to imply that through appropriate adaptation agricultural damage can on average be reduced to rather small amounts, at least for the temperature levels and regions considered. Table 2.2 provides an illustration of this point from the MINK study. It suggests that, as a result of adaptation measures, it may be possible to reduce agricultural damage to the region by almost 60 per cent (Easterling et al, 1993).

Table 2.2 Impact of adaptation on agricultural damages in the MINK region (yield changes 1984–87, incl. CO_2 enrichment)

	No Adjustment		Incl. Adjustment	
	m$ (1982)	%	m$ (1982)	%
Corn	−1035	−13.4	−1236	−16.0
Wheat	150	8.2	361	19.7
Sorghum	−118	−9.4	178	14.1
Soybeans	−438	−12.8	−48	−1.4
Hay	112	7.0	435	27.3
Total	−1329	−8.4	−532	−3.3

Source: Easterling et al (1993).

A second example is the assessment of air pollution damages which are expected to increase as a consequence of higher temperatures. Titus (1992) and Cline (1992a) both estimate the costs of additional air pollution as the extra effort required to maintain current air quality standards. In contrast, Chapter 3 below will estimate air pollution damage under the assumption that no extra measures are undertaken and air quality standards are allowed to deteriorate. The two assumptions represent two polar cases of available adaptation strategies: abate either all the additional pollution or none at all. Neither case necessarily constitutes an optimal response. Health authorities may find it desirable to adopt altogether different air quality standards in a warmer world.

---

6 See eg Easterling et al (1993), Rosenzweig and Parry (1994), and, using a somewhat different approach, Mendelsohn et al (1992).

The question of optimal policy response also dominates the literature on sea level rise. In the absence of any response even a modest rise would be disastrous. The inundation of low-lying coasts would threaten highly valuable assets and displace millions of people. Damage will, however, be much lower once appropriate protection measures are incorporated. We will illustrate this further in Chapter 5.

Worst case vs best guess

Many critics have been puzzled by the apparent dichotomy between the relatively low 2xCO$_2$ damage estimates and the potentially disastrous scenarios drawn in other parts of the literature. Catastrophe scenarios portrayed in the literature include, for example (Howarth and Monahan, 1992; Cline, 1992a):

- The melting of the polar ice caps. A possible *disintegration of the west-antarctic ice sheet* would raise sea levels by up to 6 metres. This process is, however, slow and would take place over a time span of 300–500 years.
- A *shut-down of the ocean conveyer belt* may lead to changes in ocean circulation patterns. A redirection of the gulf stream would – somewhat ironically – cause significant cooling in Western Europe, with temperatures comparable to those currently observed in Canada.[7]
- The *runaway greenhouse effect*. Initial warming levels may be amplified through massive feedback effects, eg through the liberation of methane from previously frozen sediments into the atmosphere.
- *Abrupt, non-linear changes in climate patterns*. There is paleo-climatic evidence from ice-cores pointing at the prospect of a highly unstable climate with temperature changes of several degrees Celsius within only a few years.

Evidently, the estimates of Table 2.1 do not cover extreme events of this type. They are *best guess* figures, deliberately confined to the most likely damage scenario. Given the complexity of the climatic system and the unprecedented stress imposed on it, this focus may be too narrow, though. As the above examples show, other potentially more disastrous outcomes cannot be excluded. Rather than with only one point, we are confronted with an entire damage probability distribution (see Figure 2.2). Unfortunately, only little is known about the shape of this distribution, in particular about the probability of a catastrophic outcome. As an illustration Table 2.3 reproduces some results from the Nordhaus (1993d, 1994) poll of experts which give an indication about the perceived likelihood of a climate catastrophe, i.e. about the thickness of the upper tail of the damage distribution.

In addition to expected impacts, there may also be *surprises*, events which

---

7 Abstracting from the amount of warming which will already have occurred at that date.

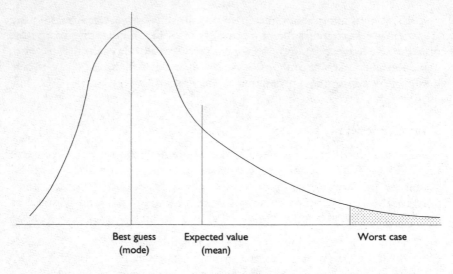

| | Best guess
(mode) | Expected value
(mean) | Worst case |

Figure 2.2   Damage distribution, catastrophic events and best guess estimates

Table 2.3   The probability of a climate catastrophe (% probability of a damage greater than 25% of GNP)

	Scenario A 3°C warming by 2090	Scenario B 6°C warming by 2175	Scenario C 6°C warming by 2090
mean value of received answers	4.8	12.1	17.5
median	0.5	3.0	5.0
trimmed mean	0.9	4.6	8.2
range of answers	0.0–30	0.2–75	0.3–95

Source: Poll of experts, Nordhaus (1993d, 1994).

are impossible to predict beforehand and for which no probability of occurrence therefore exists.

In terms of Figure 2.2, the estimates of Table 2.1 are solely concerned with the *mode* of the damage distribution. They neglect its upper tail, which is the principal theme of the catastrophe literature. Nor do they deal with surprises, for obvious reasons. By their very nature, nothing can be said about the direction or magnitude of such events. Clearly, in neither considering the entire damage distribution, nor the possibility of surprises, existing estimates are incomplete. This has to be remembered when using them to decide about the optimal greenhouse policy response. Arguably, decision-makers may be inclined to take precautionary measures to reduce the risk of a catastrophe or an unhappy surprise. We will come back to the question of the optimal greenhouse policy response in Chapter 9. Issues of damage uncertainty will also be discussed in Chapter 4.

Long-term impacts

Existing estimates are predominantly concerned with the case of $2xCO_2$. For practical purposes, however, $2xCO_2$ is only of limited relevance. In view of the long-run character of global warming it would be crucial to know more about global warming impacts beyond $2xCO_2$. Regrettably, very little work has been done in this area, both with respect to impacts in physical terms and with respect to the economic costs. The only economic assessment of long run damage costs is found in Cline (1992a), whose estimates are, however, in parts rather speculative. An attempt has also been made by Nordhaus (1993d, 1994) in his poll of impact experts. While the results on the whole confirm the 1 per cent to 2 per cent of GNP damage range for $2xCO_2$, Nordhaus observed that experts felt rather uncomfortable with respect to long run predictions. In the absence of better information, greenhouse damage costs are currently usually specified as a polynomial (typically linear to cubic) function of temperature rise, calibrated around the available $2xCO_2$ estimates.[8] Damage is further assumed to rise in proportion to GNP. Clearly, these assumptions are rather *ad hoc*, and much further effort will be required for damage assessments to reach a similar level of sophistication as has already been achieved in carbon abatement modelling.

Predicting the future

Forecasting the structure of the future societies confronted with climate change is extremely difficult. Predicting the preferences of future generations for commodities affected by warming is a near impossibility. For this reason, damage estimates will of necessity always be uncertain, even if scientific uncertainty is resolved. Enumerative estimates usually circumvent this problem by imposing $2xCO_2$ on an economy with today's structure. In most studies, base period damage is then assumed to simply grow in proportion to GNP. This is clearly an approximation and we can expect some damage aspects (eg the value of endangered ecosystems) to grow faster than others. The approach is consistent with a proposal by Cline (1992a), who advocates an initial estimation on the base of the present economic structure, with subsequent scaling to future levels.

Schelling (1992) has illustrated the difficulties of predicting future impacts by deliberating on how forecasters in the year 1900 would have predicted possible greenhouse impacts for the year 1992: 'There would have been no way to assess the impact of changing climates on air travel, electronic communication, the construction of skyscrapers, or the value of

---

8 Eg Nordhaus (1993a, b) and Peck and Teisberg (1992, 1993a, b).

Californian real estate'. However, while Schelling's observation is correct, the difficulty of the task is in itself no reason not to undertake the analysis. The current discussion about the best possible policy response suggests that a damage assessment of some sort is urgently needed, and the fact that it will be based on uncertain data should not prevent us from obtaining it. There is nothing unusual about expectations being formed subject to a limited information set. It would only be unusual not to make use of all the information available. Of course, the quality of the forecast will also play a role when deciding about the optimal greenhouse policy response. We will come back to this question in Chapter 9.

DIRECTIONS FOR FUTURE RESEARCH

Compared to other aspects of global warming economics, the economic assessment of greenhouse damage is still in its infancy, despite the fact that, in an increasing number of areas, research would be able to build on a large scientific literature. Further research is particularly required in three respects. First, the quality of existing estimates has to be improved. The emphasis has to shift away from the agriculture and sea level rise sectors, where a comparatively high level of sophistication has already been achieved, to include other damage sectors, particularly non-market related damages like climate amenity, health and morbidity, and ecosystems loss, but also the impacts on, for example, the water industry. With respect to market-related damage, more efforts will be needed to include indirect damage effects, ie to move from a partial equilibrium, enumerative approach to an integrated general equilibrium assessment.

Second, more research is needed to assess in more detail the damage in those regions which are most likely going to be the hardest hit: the developing countries. There are hardly any economic valuation studies for non-OECD countries – eg of the value of a statistical life – and such undertakings could therefore be of general methodological interest.

Third, we have to overcome the concentration on the $2xCO_2$ benchmark. The analogy approach used in the MINK study may be a possible alternative, but only to the extent that reasonable analogues are available. More research, both scientific and economic, is needed into the long-run effect of global warming and into the identification of potential thresholds. Economic damage modelling has to be raised onto the same level of sophistication as that already achieved in other areas of greenhouse economics, such as the costs of carbon abatement. Improved damage estimates will then allow for more reliable recommendations to be produced as to what the optimal global warming policy response may be. In the following three chapters some of these points will be picked up in an attempt to meet at least some of the criticisms raised.

Chapter 3

The Economic Costs of CO$_2$ Concentration Doubling

INTRODUCTION

Chapter 3 provides order of magnitude estimates of the damage associated with a doubling of atmospheric CO$_2$ concentration. Figures are produced for five different geopolitical regions as well as for the world as a whole. The estimates confirm the damage cost benchmark of 1 per cent to 2 per cent of GNP established in the previous chapter, at least for developed countries. The estimates for developing countries are about 60 per cent higher, supporting the view that the poorest countries will suffer most, even if adequate protection measures are taken.

Estimating the potential damage of global warming is a daunting task. The extent of uncertainty is vast in almost every respect. On a purely climatological level there is a fairly broad (but far from unanimous[1]) consensus about the expected changes in global mean temperature. Confidence is still low, however, in predictions of regional and seasonal variations. Difficulties also exist in predicting the changes in temperature-dependent variables such as precipitation, evaporation, storm frequency, sea level and so on. The uncertainty continues on the next level. Even if the degree of change was known exactly, the effect it would have, say, on human health, on ecosystems or on agricultural yields is far from being clearly understood.

Under these circumstances any attempt towards a monetary damage estimate can not be more than a rough assessment of the *order of magnitude*. In addition, to pursue the task several simplifying assumptions had to be introduced. The analysis is confined to an *enumerative approach* (see Chapter 2). We only analyse one particular scenario, the damage occurring with an atmospheric CO$_2$- concentration of twice the preindustrial level (2xCO$_2$). As mentioned earlier, 2xCO$_2$ is the benchmark case of almost every study dealing with the impacts of global warming. IPCC (1990a) predicts that 2xCO$_2$ will lead to an (equilibrium) increase in global mean temperature of 1.5 to 4.5°C, with a best guess of 2.5°C. Subsequent work by Wigley and Raper (1992), based on the revised IPCC (1992a) assumptions, calculates that this would induce a rise in sea level of about 50 cm by the year 2100,

---

1 See for example the critique on the IPCC findings of the Marshall Institute and others reported in Cline (1991, 1992a) and Lunde (1991).

somewhat lower than IPCC's initial prediction of 66 cm. The following results will be based on these estimates. The sea level rise figure is somewhat arbitrary, in that it does not directly relate to the $2xCO_2$ equilibrium level. Taking the year 2100 as a benchmark seems a reasonable, if somewhat conservative, approximation. Under a 'business as usual' scenario $2xCO_2$ can be expected by about 2050–60. Oceans will react only slowly, though, due to thermal inertia. By using the year 2100 as a benchmark we implicitly assume a lag time of 40–50 years.

One of the most difficult and controversial issues of greenhouse damage estimation is the prediction of future economic development (see Chapter 2). The problem is circumvented here by choosing the year 1988 as base period, ie we estimate the damage which $2xCO_2$ would cause to a world with the economic structure of 1988. For most reasonable future development paths, this method will underestimate absolute damage levels and over-estimate relative damage to developing countries. On the other hand, the approach is consistent with Cline (1992a), who advocates an initial esti-mation on the base of the present economic structure, with subsequent scaling to future levels.

Six different (partly overlapping) 'regions' are considered: the European Union, the USA, the countries of the former USSR, China, the OECD nations (including EU and US) and the World as a whole. In this respect the analysis goes beyond the US studies by Cline (1992a), Titus (1992) and Nordhaus (1991b, c), which were introduced in Chapter 2.

Climate change will affect a wide range of activities and sectors. An attempt at a classification was made earlier, in Figure 2.1. We will roughly follow this earlier categorization here and deal with each of the main aspects separately. Whenever possible, damage is expressed as the change in pro-ducer and consumer surplus or as the willingness to pay for protection. Total damage is the sum of the partial equilibrium costs in each sector. It is evaluated and discussed at the end of the chapter.

DAMAGE IN INDIVIDUAL SECTORS

Sea level rise protection costs

The rise in sea level triggered by global warming threatens to inundate vast land areas along low-lying coastlines. However, it is quite likely that at least the more valuable areas will be protected. To make use of the data from IPCC (1990c) it is assumed that only undeveloped or sparsely populated regions will be abandoned, while highly developed areas such as cities or tourist beaches will be protected (partial retreat scenario). Instead of calculating the full potential land and capital loss we therefore evaluate the combined costs of capital protection plus remaining land and capital loss. This section deals with the former aspect. The value of the unprotected dryland lost will be evaluated in the following section. It should be noted, though, that the partial retreat assumption, although certainly reasonable in many cases, is introduced mainly for data reasons and does not necessarily reflect the cost efficient response for all regions. We will come back to the question of optimal sea level rise protection in Chapter 5. 'Partial retreat' may also be quite an optimistic assumption for poorer countries, which are likely to lack the funds for sufficient coastal protection.

A major global study on the costs of coastal defence measures has been carried out for IPCC by Delft Hydraulics (IPCC, 1990c). The engineers of Delft Hydraulics estimated the world-wide costs of protecting beaches, cities, harbours and densely populated coastlines against a sea level rise of 1 metre by the year 2100.[2] The measures considered include the building of sea walls, levees and dikes, beach nourishment and the elevation of islands. They include protection as well as maintenance costs. The second column of Table 3.1 shows their estimates of the total (undiscounted) protection costs arising during a 110-year time horizon.

Table 3.1   Costs of preventing capital loss

	I m rise, Delft Hydraulics (bn$)	Annuitized costs for 50 cm rise (m$/yr)
EU	65.39	121
USA	86.82	160
Ex-USSR	24.98	46
CHINA	11.56	22
OECD	242.47	448
WORLD	495.48	915

Source: see text.

In the light of the revised IPCC findings (Wigley and Raper, 1992; IPCC, 1992a) the Delft Hydraulics assumption of a 1 metre rise by 2100 appears too pessimistic, however. In adjusting the figures we presumed a polynomial relationship between protection costs and sea level rise (Titus et al., 1991). The Titus et al estimates imply a power factor of 1.28. For a 50 cm rise the figures thus have to be multiplied by a factor of $1/2^{1.28} = 0.41$. In addition, the Delft numbers are estimates of the undiscounted total costs of protection, and therefore have to be translated into an annual expenditure stream. This is done as follows. From column 1 of Table 3.1 we deduce that world-wide investment against a 50 cm rise amounts to $495.48*0.41 = 203.15 bn over 110 years. That is $1.85 bn per year on average. Assuming the moderate

---

2 A 'densely populated' region is defined as an area with a population density of more than ten inhabitants/km².

discount rate of 1.5 per cent suggested for such projects by Cline (1992a), the present value of this annual expenditure stream is $100.67 bn.[3] The average annual protection costs are thus $915 m. Applying the same logic to the other regions yields the results shown in the last column of Table 3.1.

Dryland loss

In the previous section we introduced the assumption that all densely populated areas will be protected against the rising sea. Consequently, the loss of dryland due to climate change will be restricted to undeveloped and sparsely populated areas.

The estimated lengths of sparsely populated, low-lying coastlines in each region is shown in Table 3.2. The values are partly based on Rijsberman (1991) and partly on our own calculations. In both cases the estimation method used is that of Delft Hydraulics (IPCC, 1990c).[4] The high figure for the former USSR reflects its vast undeveloped coastal areas along the north coast of Siberia, while, at the other extreme, the figure for China is zero because its coastline is either densely populated or not 'low-lying' in the sense of the Delft Hydraulics definition. The dryland loss prediction of Titus et al (1991) under the 'partial protection' scenario implies a loss of about 0.46 km^2 per kilometre of undeveloped coastline. Assuming that this proportion is comparable to global average figures we can derive the dryland losses given in column 2 of Table 3.2.

Table 3.2 Costs of dryland loss

	Low-lying coasts (km)	Dryland loss (km^2)	Dryland loss (m$)	Lost benefits (m$/yr)
EU	3,470	1,596	3,192	319
USA	23,250	10,695	21,390	2,139
Ex-USSR	52,000	23,920	11,960	1,196
CHINA	0	0	0	0
OECD	87,870	40,420	80,840	8,084
WORLD	304,180	139,923	139,923	13,992

Source: see text.

3 $1.8468*(1 + \lambda + \lambda^2 + \ldots + \lambda^{109}) = 100.67$, where $\lambda = 1/(1 + \delta) = 0.985$ is the discount factor. See also Cline (1992a).

4 The length of coastlines with a population density of less than 10 people/km^2 and the 100 metre contour less than 25 km away from the coast are measured from the *Times Atlas* 'as the crow flies', and then multiplied by a coastline factor to allow for near-shore islands, large estuaries etc. See IPCC (1990c) for details. An exception is the value for the world as a whole, which was achieved by assuming the same ratio of densely to sparsely populated coastlines as for OECD, CHINA and ex-USSR.

Valuing coastal lands is rather difficult and figures differ by several orders of magnitude, depending on use and location of the piece of land in question. Rijsberman (1991) reports values ranging from $1 to 200 m/km². Bearing in mind that the land under threat is only sparsely populated and in general undeveloped, a rather low value seems appropriate and we therefore adopt, at least for the OECD countries, the average value of $2 m/km² used in Titus et al (1991), a value which has also been advocated by Rijsberman (1991). Such a value would, however, be too high in the case of the former USSR, where the main area under threat is the almost uninhabited north coast. For the former USSR we therefore assume an arbitrarily chosen price of $0.5 m/km². For the world as a whole we use $1 m/km², assuming that this is an acceptable average between the high price in the industrialized world and the lower prices in less developed countries.[5] Following Cline (1992a) in assuming a 10 per cent return on land per year we can derive the annual revenue losses reported in the last column of Table 3.2.[6]

Coastal wetland loss

Along with dryland and urban structures the predicted rise in sea levels will also threaten coastal wetlands, important ecosystems which are already heavily endangered by current coastal development and water drainage schemes. The possible amount of wetlands loss depends mainly on the possibility of the systems to migrate inland, and therefore on the amount of coastal protection measures taken. The more comprehensive the defence measures, the more difficult backward migration becomes and the more coastal wetlands will be lost. Titus et al (1991) estimate that for the United States a complete protection of all coastal zones would lead to a loss of one-half of all remaining wetlands. The number is lowered to one-third with only a partial protection and to 30 per cent under a complete retreat scenario.

---

5 As an example for the land price in the less developed world, Ayres and Walter (1991) report a value of $0.3 m/km² for arable land in Bangladesh. Also note their objection on using lower values for poorer countries, a line which we do not follow here.

6 Strictly, this measure is inexact. It corresponds to the area ABCE, while the true welfare loss is ABCD. Compared to our very rough estimate of land prices the error should be of second order, however.

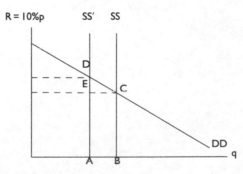

It is assumed that this estimate is comparable to world-wide values and that therefore, given the assumption of partial protection, 33 per cent of all coastal wetlands will be lost. An inventory of wetland areas in OECD countries has been provided by Rijsberman (1991). It was used to calculate the wetland loss figures in Table 3.3. The figures for the former USSR and China were derived by using the world average of wetland areas per km of low-lying, densely populated coast.[7] The services and benefits arising from coastal wetlands are manifold as is shown in the rough classification of Table 3.4. Surveys on their monetary value are found in Turner (1991) and Turner and Jones (1990).

Table 3.3 Coastal wetland loss

	Coastal wetlands lost (km²)	Value of wetlands lost (bn$)	Foregone benefits (m$/yr)
EU	9,887	49.43	4,943
USA	11,121	55.61	5,561
Ex-USSR	9,788	12.24	1,224
CHINA	11,918	5.96	596
OECD	33,862	169.31	16,931
WORLD	252,985	316.23	31,623

Note: OECD excluding Australia, Canada and New Zealand. WORLD excluding some 60 countries. The Dutch figure was corrected for a typographical error.

Source: see text.

Table 3.4 Benefits from coastal wetlands

Direct output benefits	● Commercial fishery ● Recreation (incl. sport fishing and hunting) ● Furs
Indirect functional benefits	● Flood protection ● Salinity balance mechanism ● Life support for migratory birds and fish
Non-use (existence and option) value	● Wildlife habitat ● Landscape value

Source: Turner (1991), Turner and Jones (1990).

---

7 The average was taken excluding the eight countries Argentina, Brazil, Cuba, Indonesia, Malaysia, Mexico, Vietnam and Papua New Guinea, which together harbour half of the world's coastal wetlands (Rijsberman, 1991). The remaining countries harbour 383,311 km² of wetlands along a coastline of 282,941 km (Rijsberman, 1991; IPCC, 1990c) or 1.35 km² per km of coastline.

Titus et al (1991) estimate the value of coastal wetlands as $1.5–7.5 m/ km², including all quantifiable benefits.[8] Rijsberman (1991) quotes a figure of $3 m–$13 m/km² and thus works with a median $8 m/km², while Cline (1992a) prefers a more cautious $2.5 m/km². As an overall average between these estimates we work with a value of $5 m/km² for the OECD countries. It seems reasonable to assume a different value for non-industrialized nations. In these regions the emphasis will be mainly on the returns from commercial fisheries and to a lesser extent maybe on indirect benefits, while recreational benefits are less important. For the countries of the former USSR we assume benefits of $1.25 m/km² and for China $0.5 m/km². The figure $1.25 m/km² also seems to be an appropriate average for the world as a whole, given that over 85 per cent of all wetlands are in middle or low income countries (Rijsberman, 1991).[9]

Again, assuming a return on land of 10 per cent per year yields the annual costs given in the last column of Table 3.3. Note that these estimates also include the damage to coastal fisheries, a category which was listed separately in Figure 2.1 and about which more will be said below.

Species and ecosystems loss

Most studies on the impacts of global warming predict an increased loss of species and ecosystems. Specially threatened are, according to IPCC (1990b), geographically localized and slowly reproducing species as well as poor dispersers and communities 'where climate change adds to existing stress' (p 3).

Unfortunately, more specific information is not available. Naming the species actually under threat proves difficult, not least due to the wide range of other factors which in the end determine the destiny of a species. It should be noticed, though, that some of the more prestigious mammals also fall in the categories described above (for a survey see Peters and Lovejoy, 1992).

In measuring the total value of a species or an ecosystem, economists distinguish between use, option and existence value (see eg Pearce and Turner, 1990). There are various methods to measure these, but the only method to capture all three aspects is the contingent valuation method. Pearce (1993a) reports results from several studies which yield an average willingness to pay of $9–$13 (1990) per person and year for the preservation of animals ranging from the emerald shiner to the grizzly bear. The perhaps more relevant figure for the preservation of entire habitats is somewhat higher, $9–$107 per person, or some $50 on average. A willingness to pay

---

8 At least the lower bound figure excludes aesthetic benefits. Also note that the upper bound figure coincides with rough estimates of the cost of wetland reclamation.

9 For comparison, the purchasing power parity value of per capita income in middle income countries like the former USSR is 20 per cent to 25 per cent of the OECD average, for low income countries the corresponding figure about 9 per cent, and for all non-OECD nations together 17 per cent.

estimate of about $30 (1988) per person and year in high income countries does not seem unreasonable, therefore, bearing in mind the wider, although as yet unspecified, threats from global warming.

This still rather conservative, less than average, value was chosen to account for the effect of geographical distance, which is of some importance in the present context. Unless an asset is truly unique, people's willingness to pay tends to decrease with the distance to the site in question. The effect is less frequently observed for individual species, where there are no substitutes. The $9–$13 estimate for species can therefore be seen as a lower bound, the $50 for habitats correspondingly as an upper bound. A figure of $30 per person would then seem a reasonable average. For the middle income countries (eg those of the former Soviet Union), where hardly any valuation studies are available, we assume an arbitrary value of $8 per person and year, and for low income countries like China $2/person and year.

The total cost of ecosystems loss was achieved by simply multiplying the above per capita values by the number of people living in each region. The results are shown in Table 3.5. To analyse the figures a bit further, consider the following interpretation (Cline, 1993c). Suppose there are n habitats of comparable significance to those reported in Pearce (1993a), valued at $30 per person. Adopting this value in our analysis is then equivalent to assuming that the probability of their disappearance due to $2xCO_2$ is equal to $1/n$. To get a rough impression about the size of n we can use a statistic from the World Resources Institute (1990), which reports a total of about 5000 protected parks and wildlife areas world-wide. Suppose that each person has a positive willingness to pay for only one-hundredth of these areas, ie n = 50.[10] The probability of climate-induced habitat loss implied by our analysis would be a mere 2 per cent. This then seems to suggest that our estimates are probably rather conservative.

Table 3.5 Value of lost ecosystems

	Total economic value (m$)
EU	9,750
USA	7,380
Ex-USSR	2,288
CHINA	2,176
OECD	25,470
WORLD	40,530

Source: see text.

10 Note that out of the entire set of 5000 sites each person will have a positive valuation for a different subset of n = 50. As mentioned above, people will have a preference for unique sites or those in their neighbourhood.

The figures in Table 3.5 consist of option and use as well as existence values. Note that our method then implies that people have use and existence values for the same number of sights. Evidently, this need not be the case. Existence values, on the one hand, are more or less independent of geographical locations, and people may value a variety of sites scattered around the globe. People all over the world are, for example, willing to pay for the preservation of the blue whale. Use and option values on the other hand are geographically limited, and only occur to the subset of people who actually use or profit from a certain habitat. By treating the three value categories equally, we thus introduce a bias, the size of which is however unknown.

Agriculture

Together with the costs of sea level rise the effects on agriculture are probably the most studied aspect of global warming damage. Much of this research concentrates on productivity or output aspects, though, and does not include the impact of changing prices. Price effects are crucial, however, for the economic valuation of agricultural damage. Many studies also neglect the benefits from an adaptive adjustment of the production technology (eg by using different crops etc), as reported in Chapter 2.

Our estimates are based on a study by Kane et al (1992), which includes price effects, but neglects managerial responses as well as the effect of CO_2-fertilization. Working with two scenarios, labelled A and B, Kane et al analysed the impact of climate change on crop yields, data which are then fed into a 'world food model' to analyse the effects on world agricultural markets. The model distinguishes between 13 regions and contains 22 agricultural commodities. Welfare changes (measured as changes in producer and consumer surplus) can occur in two ways: firstly by a change in a region's agricultural output due to different climate conditions, and secondly by a change in world prices.

The welfare effects (as percentage of GDP) for the two scenarios considered in the study are reproduced in the middle column of Table 3.6. Absolute values for the year 1988 are shown in the last column. They are based on average figures from the two scenarios. The results are significantly negative for all regions, but the discrepancies between the two scenarios are considerable. This is particularly the case for China, where impacts range from a loss of more than 5 per cent to benefits of over 1 per cent of GDP, and to a lesser extent for the former USSR. It should be emphasized, however, that particularly the 'upper bound' case A is quite optimistic, compared to eg Rosenzweig and Parry (1994). It assumes non-negative yield effects in most regions. Scenario B assumes negative yield effects even for northern regions such as Canada and the former USSR, and is more within the Rosenzweig and Parry range.

Climate change may aggravate the world's hunger problem. Rosenzweig

Table 3.6 Costs to agriculture

	Range of welfare change (% GDP[a])	Average welfare change (m$[b])
EU	−0.400 .. −0.019	−9,666
USA	−0.310 .. +0.005	−7,392
Ex-USSR	−0.520 .. +0.032	−6,185
CHINA	−5.480 .. +1.280	−7,812
OECD	−0.316 .. −0.018[c]	−23,130
WORLD	−0.470 .. +0.010	−39,141

a Range from the two scenarios A and B of Kane et al (1992).
b For ex-USSR the result is based on GNP rather than GDP.
c Average over several subregions.

Source: compiled from Kane et al (1992).

and Parry (1994) predict a significant increase in the number of people at risk of hunger. In their most plausible scenarios including managerial responses by farmers, predictions range from a slight decrease in the number of people at risk to an increase of up to 50 per cent, for different climate scenarios (see also Rosenzweig et al, 1993). Taking the average and adjusting for a lower warming rate, this translates into an increase of about 10 per cent for 2.5°C. Arguably, this number could be significantly reduced through appropriate precautionary measures, though. Rather than causing a redirection of funds away from development projects, climate change could thus actually strengthen the case of enhanced development aid, which reduces climate vulnerability as well as being desirable on moral grounds.

Forestry

Roughly one-third of the world's land surface is covered by forests or woodlands (IPCC, 1990b). The extent to which this area will be affected by climate change depends on various factors like, for example, the species and age of trees, possibilities for forests to migrate and the quality of forest management. The impact of global warming on wood production is therefore ambiguous. The IPCC (1990b) assumes that although stand growth rates may increase in some areas, the overall net increment (including mortality) will be negative. Regional impacts will be strongly influenced by the extent to which forest zones can shift northwards.

Sedjo and Solomon (1989) used the Holdridge Life Zone classification to estimate that as a consequence of $2xCO_2$ the world-wide forest area could reduce by about 6 per cent. Temperate and boreal forests would decline more, by about 16 per cent, whereas tropical forest areas would expand by some 9 per cent. These figures form the basis of the estimates in Table 3.7.

Table 3.7 Damage to the forestry sector

	Loss in forest area (100 km^2)	Forestry loss (m$)
EU	52	104
USA	282	564
Ex-USSR	908	363
CHINA	121	24
OECD	901	1,801
WORLD	1,235	2,005
temperate	2,169	2,284
tropical	−934	−279

Source: see text.

Adjusted to 2.5°C warming they correspond to changes of −3.5 per cent, −9.6 per cent and 5.2 per cent, respectively. Further underlying assumptions are as follows. Forest and woodland area statistics can be found in FAO (1991a). Based on Sedjo and Solomon and World Resources Institute (1992) it was assumed that 40 per cent of all forest areas are tropical, and that no tropical forests can be found in the OECD, the ex-USSR, and China. That is, in the five regions EU, US, China, ex-USSR and OECD forest areas uniformly decrease by 9.6 per cent. For the world as a whole the picture is mixed. Forty per cent of all forest areas are growing – those in the tropics – the remaining 60 per cent are decreasing. Table 3.7 shows the reduction in forest areas implied by these assumptions.

The value of forests was estimated by Titus (1992) as $11,000–$37,000/km^2, based on observed differences in land values before and after logging. This figure is roughly in line with the ratio of income from the forest sector relative to forest area in countries with comparatively small forest areas like Germany or France. It is, however, more than an order of magnitude too large for a country with wide forest areas, such as Canada. Unfortunately most countries do not report forestry income in their national income statistics. Yet, a very small number of countries do so, and from there we deduced an average forest value of about $2,000/km^2.[11] The value in middle income countries is assumed to be $400/km^2 and in low income countries $200/km^2. The resulting forestry loss is also reported in Table 3.7.

These figures are inexact in several ways. First, they are based on an equilibrium assessment of $2xCO_2$ damage, ie after enough time has passed for forests to migrate or adjust. It has been pointed out by Cline (1992a) that

---

11 Canada, France, Italy and Germany together receive an income of $7.6 bn from the forestry sector (UN, 1990), about 0.25 per cent of their total GDP. They have an area of 3.87 million km^2 covered by forests (FAO, 1991a).

the slow adjustment speed of forest systems may cause a temporary decline in forested area over 200–300 years, before a new equilibrium is reached. Our estimates would thus be too optimistic. On the other hand they neglect the managerial response from the forestry industry, which may help to ease both transitional and equilibrium losses. On a conceptual level it should be noted that the approach used here is only an approximation of the exact welfare changes to producers and consumers. A more accurate analysis would have to be based on a general equilibrium assessment which allows for price changes as well as trade effects.[12] Finally, it should be noted that the valuation is restricted to timber benefits, ie it neglects non-timber aspects like, for example, the aesthetic or recreational value of forests. To some extent these are already included in the figures on ecosystems loss.

Fisheries

As one of a few sectors, the fishing industry will be affected by both the rise in sea level and the changing climate itself. Of the coastal infrastructure threatened by sea level rise a large proportion can be associated to fisheries. Changing climate patterns will affect the location and quality of fish grounds, as species move to new grounds or, in the worst case, simply disappear. Of particular importance for the fishing industry could be the loss of coastal wetlands. Wetlands serve as habitat or breeding ground for various species and, through the food chain, changes in this area could easily spread. Bigford (1991) estimates that a 50 per cent reduction in marsh productivity (for whatever reason) would lead to a 15 per cent to 20 per cent loss in estuarine-dependent fish harvests. Given an expected loss of about 33 per cent of all coastal wetlands (see above), we can expect a loss of 10 per cent to 13 per cent in estuarine-dependent fish harvests. Bigford also estimates that about 68 per cent (by weight) of all commercially harvested species in the US are in some way estuarine dependent. This would then imply a reduction in total catches of 7 per cent to 9 per cent in the US. Assuming that this average holds world-wide we derive the reductions in annual catches shown in Table 3.8.

Remember, however, that the estimates for wetland loss already include the damage to commercial fisheries. The figures of Table 3.8 are thus only for illustration. Including them in the total damage costs would lead to double counting.

Energy

In addition to being the target of most global warming prevention policies, the energy sector will face a significant shift in the demand for space heating

---

12 Such an analysis has been carried out by Binkley (1988). Unfortunately the study is restricted to boreal forests and is based on a rather optimistic damage scenario which assumes an overall increase in forested area.

Table 3.8 Reduction in fish harvests

	Nominal catches (1988, 1000 t)	Reduction (8%, 1000 t)
EU	6,977	558
USA	5,656	452
Ex-USSR	10,171	814
CHINA	5,806	464
OECD	31,288	2,503
WORLD	85,358	6,829

Source: see text; Nominal catches from FAO (1991b).

and cooling. There will also be effects on the energy supply, particularly through changes in the availability of fuelwood and water for hydropower generation. Unfortunately, an estimation of the former effect is not possible with the present data. The latter are included in the assessment on water damage below. The present section concentrates on demand effects.

The US Environmental Protection Agency (EPA) predicts a climate-induced increase in electricity demand of about 1.4 per cent by 2010 (for 1.2°C warming) and 5.2 per cent by 2055 (for 3.7°C), or some 1 per cent to 1.5 per cent per °C (see Smith and Tirpak, 1989; and Cline, 1992a). Adjusted to our assumption of 2.5°C warming, the EPA estimates imply an average increase in US electricity demand of about 3.2 per cent for 2xCO$_2$.

The regional differences are, however, considerable. The 2055 estimates, for example, range from a moderate decrease in the northern and north-eastern states to increases of up to 10 per cent to 15 per cent in the south. Transferring the US average to other regions is therefore dangerous. Nevertheless, on a rough and ready basis, it can be argued that the US climate mix may be roughly representative for at least the OECD, the EU and to a lesser extent also for China and the world as a whole. It is clearly not applicable for the ex-USSR, though. For this region we assume a value of –1 per cent (ie a reduction in electricity demand), based on EPA's regional estimates for the US north and north-east. The approximate increase in energy costs derived from this method is shown in Table 3.9.

The approach simplifies in at least two ways. First, the EPA study is restricted to electricity demand and neglects other forms of energy such as fossil fuels. For the US it is assumed that the demand for non-electricity energy could fall, since a large share of fossil fuels is typically used for space heating (Nordhaus, 1991b, c; Cline, 1992a). The limitation to electricity may thus lead to an overestimation of the total expenditure increase. Second, we assume constant prices. Thus, we neglect the capital costs and possible price rises which could occur if the shift in demand necessitates capacity expansions. This would imply a downward bias in our results. For

Table 3.9 Costs of increased electricity demand

	Electricity demand (1988, TWh)	Electricity price (1988, m$/TWh)	Increased energy costs (m$)
EU	1,693.7	129	6,992
USA	2,874.8	75	6,900
Ex-USSR	1,705.0	40	−682
CHINA	545.2	40	698
OECD	6,601.1	97	20,490
WORLD	11,061.4	n.a.	23,065

Source: see text. Electricity demand from IEA (1991) and OECD (1991). Prices compiled from IEA (1992) and World Bank (1992).

the US, for example, Cline (1992a) has estimated capital costs in the order of $500 m annually.

It is important to note the strong connection of the costs of increased energy demand with the amenity value of climate. Heating or cooling expenditures are adaptation or defence measures, similar in character to sea level rise protection activities. They are made in order to avoid climate disamenities, by adjusting the (inside) temperature to a more favourable level. Much as in the case of sea level rise, the total amenity costs then consist of two aspects: the costs of defence – considered here; plus the costs of the remaining temperature disamenity, with which we will deal in a separate section below.

Water

Global warming will affect both the supply and demand for fresh water. Higher temperatures are likely to cause an increase in water demand. At the same time the supply of water will be affected, mainly through the change in precipitation patterns and, in coastal areas, through the intrusion of saline water into freshwater reservoirs. The damage from salt-water intrusion is largely unknown. A Dutch study quoted by Rijsberman (1991) estimates salinity damage for Holland at $6 m a year, but wider studies are not available.

Abstracting from groundwater and other reservoirs, the amount of water available in a certain period of time is, roughly, the difference between precipitation and evapotranspiration in that period. Both factors will be influenced by global warming. Higher temperatures will lead to faster evaporation, which in turn will cause more precipitation, as the capacity of the atmosphere to store water is only modest. Global estimates predict an increase in precipitation of 7 per cent to 15 per cent and one in evapotranspiration of 5 per cent to 10 per cent. The annual run-off would thus increase on average (see Schneider et al, 1990). The confidence in these estimates is, however, low. Further, seasonal and regional differences will be considerable and many regions will face a lower run-off during at least some parts of the year. According to Schneider et al these include the American Midwest, mid-Europe, South Canada and probably also parts of Siberia and South China.

Following Cline (1992a) the welfare loss from a reduced water supply is approximated by the monetary value of the quantity cut back.[13] Based on an EPA study on Southern California, which predicts a 7 per cent to 16 per cent reduction in annual water resources, Cline assumes a 10 per cent reduction in water availability for the United States as a whole. Bearing in mind that Southern California is part of a zone which will probably be hit at a higher than average level, we prefer to work with the lower bound value of 7 per cent loss in each region. Figures on annual water withdrawals can be found in the data tables of the World Resources Institute (1990). They are reproduced in Table 3.10 (latest available year).

The water prices in each region are derived from *The Economist* (1991). The short statistic given there shows 1991 water prices for 13 OECD countries, with an average price of 56 cents/m^3 (weighted by withdrawal, adjusted to 1988). The EU average is 92 cents/m^3 (based on seven countries), while the US price is 42 cents/m^3. These values are rather high, compared to Cline (1992a) and Titus (1992) who use values of 8–20 cents/m^3 for the US. The discrepancy could be due to differences between water prices in urban areas and for agricultural use. Even accounting for this, the Cline and Titus values appear to be somewhat low, however, as Cline

---

13 The relation between the true welfare change and the approximation is shown in the following graph:

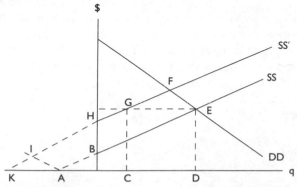

The reduction in supply (leftwards shift of the supply schedule from SS to SS′) reduces welfare (consumer and producer surplus) by the area BEFH. The approximation is CDEG. The difference between the two values is ABHI, as CDEG = KAEG = AEFI = ABHI + BEFH. Ie in the case shown here (with a positive intercept term in the supply function) the true loss is overestimated.

Table 3.10 Damage to the water sector

	Annual water withdrawal (km³)	Welfare loss in water sector (m$)
EU	218	14,039
USA	467	13,730
Ex-USSR	353	2,965
CHINA	460	1,610
OECD	889	34,849
WORLD	3,296	46,749

Source: see text.

(1993c) has himself conceded. Our estimates are based on the figures from *The Economist*. In addition we assume for the former USSR and other middle income countries a price of 12 cents/m³, and for low income countries like China 5 cents/m³.

The loss to the water sector under these assumptions is shown in Table 3.10. It should be emphasized again that these are averages. For areas in which run-off will increase and for those with an abundant supply of water the figures may be too high. For many arid and semi-arid zones, however, a further decrease in the supply of an already scarce commodity could be disastrous. The fact that areas of both types can be found in each of the regions considered gives some credibility to the average values in Table 3.10.

Services are not only taken from water withdrawals, the only aspect considered here. Titus (1992) has also estimated the costs of increased water pollution. A reduced stream flow would mean pollutants would be carried away more slowly, requiring additional clean-up efforts to maintain standards. For the US, Titus estimates these to be in the order of $33 bn for a 4°C temperature rise. Instream uses, eg from recreation or fishery, may also be significant, albeit difficult to assess.[14]

Other sectors

Generally, every sector which in some way depends on climate will be affected by global warming. Abstracting from impacts through second round effects, the areas usually highlighted in addition to those already dealt with are construction, transport, tourism and the insurance industry.

14 To give an impression of the order of magnitude, Hervik et al (1987) estimated the willingness to pay of Norwegian households to prevent rivers from being subject to hydroelectric development as 800–1600 NOK (ca. $120–240). Further valuation studies, mainly on water *quality*, are reviewed in Pearce et al (1992).

Unfortunately, data for a monetary valuation are not available for either sector.

The construction and transport sectors are both affected by cold weather, snow and ice, and both sectors might thus profit from a warmer climate, particularly through reduced disruptions and lower winter maintenance costs (see eg UK Climate Change Impact Review Group, 1991; IPCC, 1990b). On the other hand, as is pointed out by Cline (1992a), heatwaves can be similarly disruptive, in the case of transport, for example, through heat stress on railway tracks. Probably more importantly, both sectors are also negatively affected by rainfall, and precipitation is likely to increase under $2xCO_2$. IPCC (1990b) also predicts disruptions through the melting of permafrost soils and, due to changes in water levels, for inland shipping. In general, however, IPCC expects the impacts on the transport sector to be quite modest. The same is probably true for the construction sector.

For tourism and leisure, IPCC (1990b) expects an expansion of summer recreation activities mainly at the expense of the skiing industry. It has been estimated, for example, that as a consequence of a shortened period of snow cover Quebec would lose 40 per cent to 70 per cent of its skiable days (Canadian Climate Program Board, 1988a). In Ontario, the shorter skiing season may cause a loss of up to $50 m of revenue, which would, however, be offset by an increase in the camping season of up to 40 days in some areas (Canadian Climate Program Board, 1988b; Nordhaus, 1991d). Summer tourism would generally benefit from a longer season, although, as Cline points out, only to the extent that activities are not hindered by excessive heat or increased rainfall.[15] A monetary estimate for the US has been obtained by Cline (1992a), who predicts a loss from leisure activities in the order of $1.7 bn per year. However, the figure is based solely on the costs of forgone skiing days and neglects possible gains in summer activities. It may therefore be too pessimistic.

The insurance industry will mainly be affected through an increase in extreme weather events such as droughts, floods and storms. A rough estimate for increased hurricane damages will be given in a separate section below.

Human amenity

It is hardly disputed that climate is an important factor of the quality of life. Global warming will therefore also affect human amenity. It has been claimed that this effect could be beneficial, given that warmer weather is in general preferred to cooler. However, warmer weather is not better throughout. There seems to be an optimal temperature level, beyond which

---

15 Note that the impacts of the protection or loss of tourist beaches are already included in the estimates for dryland loss and sea level rise protection. Also, the impacts associated with coral reef death, emphasized by Cline (1992a), are, as a forgone use value, included in the figure for ecosystems loss.

further increases are detrimental. Mearns et al (1984), for example, predict that for a place like Des Moines the occurrence of heatwaves – temperatures above 35°C for at least five consecutive days – may increase by a factor 2–6 (for 1.7°C warming and a summer mean of 28 to 31°C). Arguably, not everybody will consider such a prospect as beneficial. The overall effect of global warming on human amenity is thus ambiguous, the impact being positive in colder and negative in warmer regions.

The case of a warmer area is depicted in Figure 3.1. An increase in temperature, ie a shift in the endowment from E to E′, makes people move from equilibrium O to O′. In the new optimum O′ utility has decreased from U_2 to U_3, cooling expenditures have increased by AB. Note that in the absence of defence measures utility would have fallen to an even lower level U4 (not shown in the graph), going through point F. The willingness to accept the welfare loss U_2–U_3 is yC. In principle, it is this distance we are

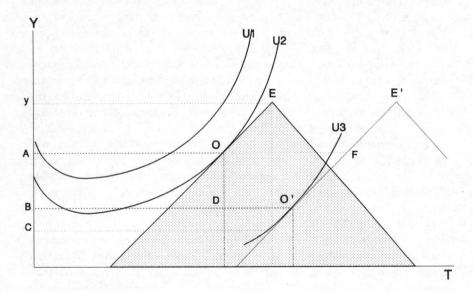

Suppose individuals gain utility from two goods, income Y and (inside) temperature T. The indifference curves are U-shaped, reflecting the fact that there is an optimal temperature level, after which further temperature increases lead to welfare loss. Individuals are endowed with an initial income and temperature bundle E. Income can either be consumed or spent on heating and/or cooling. Heating corresponds to a move downwards-right (a warmer temperature is substituted for income) and cooling to one downwards-left. The shaded triangle thus represents the feasible set or budget constraint, with the slopes of the right and left leg determined by the price of heating and cooling, respectively. The graph shows the case of a temperature beyond the optimal level, and the individual thus spends yA on cooling (optimal point O, with utility level U_2). Global warming leads to an increase in temperature, ie to a rightwards shift of E to the new endowment E′. In the new optimum O′ more money is spent on cooling and utility is at the lower level U_3. Global warming has led to a welfare loss of U_2-U_3, and an increase in cooling expenditure (energy demand) of AB.

Figure 3.1   Global warming, energy demand and welfare

interested in when aiming at a valuation of the amenity costs (or benefits) of climate change. It is possible to split yC into two parts. First, people need to be compensated for the additional cooling expenditures they undertake in order to mitigate the most adverse effects – the distance AB in Figure 3.1. Second, they need to be compensated for the remaining, unmitigated temperature increase DO′. To avoid double counting, we should in this section only be interested in the second aspect, ie the distance yC – AB. The costs of adaptation have already been assessed in the section on energy.

Unfortunately, estimating this remaining area proved impossible with the currently available information. The monetary value of a benign climate is still largely unknown, although attempts towards a valuation can be found in Hoch and Drake (1974) and Gyourko and Tracy (1991). A careful distinction between winning and losing regions would further require fairly accurate information about regional and seasonal temperature patterns.

Morbidity and mortality

Human beings are very capable of adjusting to climatic variations and, as opposed to most other species, can live in more or less every climate on earth. Nevertheless, climate change will have its impacts on human morbidity and mortality. The literature on the health effects of global warming (eg Haines and Fuchs, 1991; IPCC, 1990b; Weihe and Mertens, 1991; WHO, 1990) predicts an increase in climate-related illnesses such as cardiovascular, cerebrovascular and respiratory diseases. Summer mortality from coronary heart disease and strokes may increase and is likely to offset the reduction in winter mortality. Particularly at risk are individuals with 'immature regulatory systems, such as infants and children, and [...] those with reduced adaptive regulatory functions, such as the elderly and physically handicapped' (WHO, 1990, p 17).

In addition to these direct effects there may be changes in the occurrence of communicable diseases and an aggravation of air pollution (see the following section). The risk areas of communicable diseases like malaria or yellow fever may shift as their vectors adjust to new climate conditions. WHO (1990) fears that, for example, the so far disease-free highlands of Ethiopia, Indonesia and Kenya may be invaded by vectors. Although vector-borne diseases occur predominantly in developing countries, impacts need not be restricted to these areas. A spreading of vector-borne diseases has, for example, also been predicted for the United States and Australia (see eg Haines and Fuchs, 1991; and Smith and Tirpak, 1989). Much will, of course, depend on adaptation and the precautionary measures taken. Although not cost free, precautionary action is likely to be a major aspect of the cost-efficient response. Many projects, like improvements in the hygiene standards of poor countries, are more than likely to be worthwhile in their own right. For a proper assessment of the morbidity damage, however, much more research is still required.

Some, albeit controversial, estimates exist on the warming-induced change in mortality. In a case study carried out for EPA, Kalkstein estimated the change in mortality in 15 American cities (see Smith and Tirpak, 1989; Kalkstein, 1989). He concluded that without acclimatization the number of summer deaths in the observed area would rise by 6,246 in absolute terms, corresponding to a further 294 casualties per million inhabitants. Winter mortality, on the other hand, would decrease by nine deaths per million. The estimates were achieved by feeding $2xCO_2$ climate data into a statistical model regressing mortality on climate variables.

In a second estimate Kalkstein uses analogues to estimate the impacts under full acclimatization, including biological and behavioural adjustments as well as changes in the physical structure of a city (different architectonic style, more reflecting materials, etc). With acclimatization the increase in summer mortality falls to 49 deaths per million, and the decrease in winter mortality becomes four deaths per million. Because of the different methods used, the two pairs of estimates are not directly comparable and are partly contradictory (eg one would expect the decrease in winter mortality to be higher with acclimatization than without). The detailed Kalkstein figures underline the importance of acclimatization, showing that cities already accustomed to a warmer climate are far less affected by a further warming than cities with a moderate climate.[16] Whether, and if so to what extent and how quickly, a society will acclimatize is, however, a fiercely debated issue. It is mainly for this matter that the climate impacts on mortality remain a controversial question (see Kalkstein, 1989; and Cline, 1992a).

Nevertheless, we will take Kalkstein's more moderate estimate including acclimatization as a basis for a rough mortality estimate. Adjusted to 2.5°C it corresponds to a net increase of 27 deaths per million people. This figure may still be rather on the high side. Smith (1992), for example, reports an annual average of about 500 heat-related deaths in the US between 1936 and 1975. Following Mearns et al (1984) in assuming a two to sixfold increase in heatwaves, this would imply an increased mortality of at most 17 deaths per million (assuming an average population of 180 million between 1936 and 1975; after World Resources Institute, 1990). The discrepancy may partially be explained by the fact that Kalkstein's estimates are entirely based on urban areas, which tend to pose a greater risk (Smith, 1992). Although the 15 city studies exhibit some differences according to climate zones, the average net mortality increase of 27 deaths per million is assumed to hold in each region. The increase in mortality for each region is shown in Table 3.11.

To arrive at a monetary valuation the number of casualties has to be weighted by the value of each life. It seems worthwhile to elaborate a bit

---

16 This to an extent that some of Kalkstein's estimates actually show a *decrease* in summer mortality, the negative effects from warming being more than offset by the better acclimatization in the chosen analogue.

Table 3.11 Damage from increased mortality

	Increased mortality (deaths)	Loss from increased mortality (m$)
EU	8,775	13,163
USA	6,642	9,963
Ex-USSR	7,722	2,317
CHINA	29,376	2,938
OECD	22,923	34,385
WORLD	137,727	49,182

Source: see text.

further on this potentially controversial issue. A first crucial point to note is that we do not attempt to measure the value of an (individual) life as such. There is no such figure. What is measured is the amount of money needed to compensate people for an additional risk of death, that is the value of safety. Multiplied by the change in the risk of death, this yields a figure which is often called the value of a *statistical* life. For example, if in a town with 50,000 inhabitants residents are willing to accept $15 to tolerate an increased risk of death of 1/50,000, then the increased risk results in one statistical death. The value of that statistical life is then $15 × 50,000 = $750,000.

Various methods exist to estimate the value of a statistical life. The main distinction is between the human capital or gross output approach, which values the net present value of lost future output, and the willingness to pay approach, which estimates people's willingness to pay for increased safety. Estimates based on the latter method, all for developed countries, range from about $200,000 up to over $16 m, with an average of around $3 m (Pearce et al, 1991; Viscusi, 1993). The results suggest that a statistical life should not be valued at less than $700,000 and should plausibly be at least $1.5 m (Pearce et al, 1991). For developed regions we thus assume $1.5 m, still a fairly conservative value. A low value is, however, preferred to counterbalance the rather high quantity estimate.

People's valuation of the good safety is case-specific and depends on a variety of factors, such as the type of hazard encountered. Some risks are more acceptable than others, eg those connected to leisure activities. Values also depend on people's income level, in the same way as their willingness to pay for market goods does. We should therefore expect the value of a statistical life in poorer countries to be lower than in rich nations. This of course does *not* mean that the life of, say, a Chinese is worth less than that of a European. It merely reflects the fact that the *willingness to pay* for increased safety (a lower mortality risk) is higher in developed countries.

Unfortunately few value-of-life studies exist concerning the less developed world. We therefore used an arbitrary value of $300,000 for middle income and $100,000 for low income countries. The willingness to pay in each region to prevent an increase in mortality is given in column 2 of Table 3.11.

Air pollution

Given the wide concern about air quality and air pollution it is surprising how little attention this aspect has gained in the context of global warming so far. Global warming will affect the quality of the air in two ways.

The first has to do with what is called *secondary benefits*. Because no economical CO_2-removal technologies currently exist, all attempts to limit CO_2 emissions concentrate on cutting down the use of fossil fuels. All CO_2 abatement will, therefore, also lead to a reduction in the emission of major pollutants such as SO_2, CO and NO_x (Ayres and Walter, 1991; Pearce, 1992). However, since secondary benefits are related to *emissions abatement*, rather than *atmospheric concentration levels*, they are not relevant in the present context. We prefer to interpret them as a (negative) part of the abatement costs, and will come back to them in Chapter 6.

Many chemical reactions also depend on temperature, and this is the second way in which global warming will affect air quality. Scientists predict a warming-induced increase in the emissions of hydrocarbons (HC), nitrogen oxides (NO_x) and sulphur oxides (SO_x). In addition the formation of acidic materials could increase. The effect on acid depositions is nevertheless unclear, because of changes in clouds, winds and precipitation. More certain is an increase in the tropospheric ozone level, brought about through the increase in NO_x and HC emissions as well as through a higher reaction rate (see Smith and Tirpak, 1989). Two case studies carried out for the US EPA predict a change in O_3 concentration of -2.4 to 20 per cent. As a consequence the exposure to concentrations above the US threshold of 120 ppb (measured in people-hours) would increase by 60 to 210 per cent (Smith and Tirpak, 1989). Adjusted from the assumed 4°C warming to 2.5°C we can thus expect an average increase of 5.5 per cent in ozone concentration and 80 per cent in exposure.

For simplicity we suppose that the increase in O_3 concentration is fully due to the increase in NO_x and HC emissions, ie we assume an increase in NO_x emissions of 5.5 per cent. This may look like an overestimate, given that part of the O_3 increase is caused by a higher reaction speed. To achieve a monetary estimate the assumption seems appropriate, however, as the damage from an increased O_3 concentration is fully attributed to NO_x in all available estimates. For SO_2 we assume a rise in emissions of 2 per cent. The emissions increase in absolute terms is shown in the first two columns of Table 3.12, based on the regional estimates of the World Resources Institute (1990).

The monetary value of air pollution damage has been estimated by several authors, including Alfsen et al (1992), PACE (1990) and Pearce (1994b). The estimates for NO_x range from about $0.10 to $15 of damage per kg

Table 3.12 Damage through increased air pollution

	Increase in NO$_x$ emissions (1000 tons)	Increase in SO$_2$ emissions (1000 tons)	Damage from increased air pollution (m$)
EU	566	285	3,543
USA	1,073	422	6,420
Ex-USSR	1,584	1,100	2,134
CHINA	227	258	178
OECD	1,943	873	11,898
WORLD	4,545	2,737	15,402

Source: see text. Emissions figures are based on World Resources Institute (1990). Their 1982–84 estimates for 21 OECD countries, Eastern Europe and China were taken unadjusted as there hardly is an overall time trend. For non-reporting regions estimates were derived from the emissions/GNP ratios of reporting countries as follows: for OECD countries we used the ratio of reporting OECD regions; for the former USSR and Eastern Europe we used that of reporting Eastern European countries; all other regions are based on the overall ratio of all reporting regions.

emitted. The figure is exclusive of the damage from acid rain, which was subtracted because its affectedness by global warming is as yet unclear.[17] The divergence in the figures mainly stems from differences in the assessment of health impacts, which account for most of the damage in the higher Alfsen et al estimate, but are assumed zero in the figure by Pearce. In the following we use an average of $5/kg for developed countries, $1/kg for middle income countries and $0.5/kg in LDCs.

Again excluding acid deposition, SO$_2$ causes a damage of $0.6/kg to 3.5/kg, mainly through health effects and corrosion. We use an average value of $2.5/kg in OECD countries. For middle and low income countries we assume $0.5/kg and $0.25/kg, respectively. The estimates resulting under these assumptions are shown in Table 3.12.

Migration

Global warming could trigger a large migration stream away from the worst-affected regions. Ayres and Walter (1991), for example, assume that as many as 100 million people could be displaced world-wide. Cline (1992a) quotes estimates of 72 million people displaced in China, eight million in Egypt and half-a-million in Poland. Ali and Haq (1990) estimate 11 per cent of total population could be affected in Bangladesh (*ca* 11.4 million people).

However, all these studies, which concentrate on displacements due to the rising sea, assume a worst-case scenario in which no protection measures are

---

17 A clear separation of acid rain and pure NO$_x$ damage was not always possible. It should also be noted that the estimates are strongly site-dependent.

taken at all. The view in this study is that, as a cost efficient response to sea level rise, densely populated coastlines will be protected (see the section on protection costs, above). Under this assumption the number of people displaced will of course be lower. Climate-induced migration may nevertheless still occur, eg away from unprotected coasts or from regions where climate became unfavourable for agriculture. The type of migration will range from voluntary resettlements to the occurrence of actual climate refugees, where the former group will cause the least (if any) costs and the latter probably the highest, specially if non-economic disutilities (eg from stress and hardship) are included.

Cline (1992a) guesstimates that global warming will lead to 100,000 additional immigrants per year for the US. This corresponds to an increase of roughly 17 per cent compared to 1988 long-term immigration (UN, 1991). For lack of better data we assume that this percentage is representative for the world as a whole. Immigration figures for the OECD countries were taken from UN (1991).[18] The OECD ratio of additional immigrants per capita was then used to derive estimates for the remaining regions. Using the high OECD ratio probably leads to an overestimation of the non-OECD figures. On the other hand, we have neglected domestic migration, which may be significant in regions like China and the former USSR which stretch over different climate zones. The estimated immigration increases are shown in Table 3.13.

Table 3.13 Migration costs

	Additional immigrants (in 1000)	Total costs (m$)
EU	229	1,031
USA	100	450
Ex-USSR	153	153
CHINA	583	583
OECD	455	2,048
WORLD	2,734	4,327[a]

a 455,000 immigrants to OECD countries at $4,500, remaining immigration at $1,000.

Source: see text.

The costs of increased migration are estimated in Cline (1992a) and Ayres and Walter (1991). Despite using completely different methods both studies come up with an estimate of roughly $4500/immigrant for the United States.

18 Note that they include short-term immigration for some countries.

Although neither method is fully convincing, this value is used to estimate the immigration costs in OECD countries. For poorer countries Ayres and Walter assume costs of $1000 per person. This value is deduced from the forgone output a person would have produced, had he or she not migrated. The mere costs of resettlement and support of a climate refugee are much lower and would only be $72 per person and year (*ibid*). This latter figure is based on the average expenditures of the UN High Commissioner for Refugees and the World Food Programme. The $1,000 estimate is used for all immigrants to non-OECD regions.

To these costs would have to be added the costs of hardship and stress suffered by migrants. As Cline puts it, 'peoples have often fought wars to avoid being forced to leave their homelands' (1992a, p119), and it is therefore quite likely that these costs exceed the pure economic losses. Unfortunately, it seems almost impossible to assess them properly. For a conceptual discussion see Pearce (1993b).

Natural disasters

Under $2xCO_2$ extreme events like floods and droughts are likely to become more frequent. IPCC (1990a) also predicts, albeit with only a low confidence, an increase in local rainstorms at the expense of gentler but more persistent rainfalls. Tropical storms (hurricanes, typhoons) may become more frequent and wider spread and could occur with increased intensity. Mid-latitude winter storms may diminish, while the Asian summer monsoon could intensify. Such changes in occurrence and intensity of natural hazards could be connected with considerable costs in terms of both casualties and destruction. Already today more than 45,000 people die each year as a consequence of natural disasters (including seismic hazards), more than half of them in tropical cyclones (Bryant, 1991). As no other estimates exist, this section concentrates on the damage from tropical cyclones. The total damage from increased natural hazards is therefore likely to exceed the estimate given below.[19]

Cyclones can only form over warm oceans with sea surface temperatures above 26°C. Consequently, they only occur in certain areas, the most important being the South West Pacific, Eastern Asia and the Caribbean Sea (see Figure 3.2). Judging from the length of coastline affected, and taking into account the different frequencies of occurrence, we deduced the following rough occurrence estimates. The United States is affected by 6.6 per cent of all cyclones, some 7.2 per cent affect China, and 28.9 per cent occur in OECD nations (Australia, Japan, New Zealand and the US). Of all storms, 0.4 per cent reach as far north as to affect the ex-Soviet Union.

---

19 Note, however, that some of the damage from increased sea-flooding is averted through the protection measures discussed earlier. (Not prevented is the flooding of unprotected zones.) Also, it is not entirely clear as to how far the EPA study on California, on which the water damage estimate is based, includes droughts.

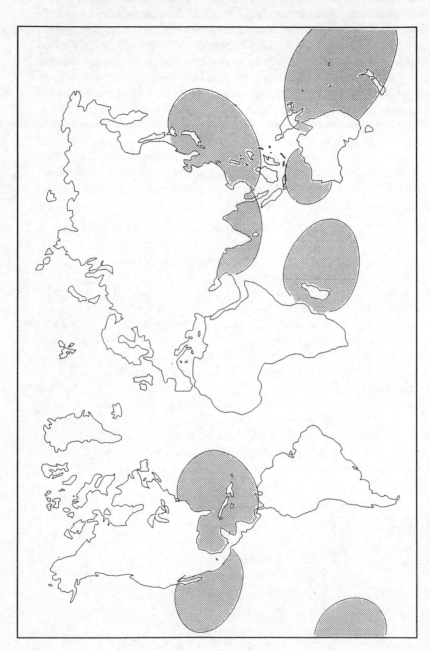

Source: after Berz (1990), Bryant (1991) and Smith (1992).

Figure 3.2 Occurrence of tropical cyclones

Neglecting overseas dominions, tropical storms are unknown in EU countries.[20]

In an average year, about 70 to 80 tropical cyclones are recorded world-wide. The annual damages have been estimated at about $1.5 bn, with a death toll of 15,000 to 23,000 lives (Smith, 1992; Bryant, 1991). The average values are overshadowed by the disastrous impacts of extreme events, though. In 1970 a cyclone caused more than half-a-million deaths in what is now Bangladesh. In 1985 the disaster was repeated, killing another 100,000 people (Bryant, 1991). Between 1988 and 1992 the three hurricanes Gilbert, Hugo and Andrew caused damages of several billion dollars each. In the light of these considerations the average figures appear to be rather on the low side. On the other hand, the probability of occurrence of an extreme event like hurricane Andrew has to be accounted for as well.

The impact of global warming on tropical storms has been analysed by Emanuel (1987). He estimates that $2xCO_2$ could lead to an increase in the destructive power of tropical storms of 40 per cent to 50 per cent, for 4.2°C warming. Adjusted to 2.5°C warming, this is about 28 per cent. It seems reasonable to assume that storm damages rise more than linearly with intensity. Suppose they rise to the power 1.5. An increase in destructive power of 28 per cent would then increase damage by $28 \times 1.5 = 42$ per cent.

Accepting the estimates by Smith and Bryant, this would imply an additional $630 m in damages and about 8000 more lives lost. Again, these figures appear rather low, and may arguably constitute a lower bound. The distress and mental health effects suffered by survivors certainly remain uncounted. Studies with survivors found traumatic neurotic reactions in 80 per cent of the cases (Haines and Fuchs, 1991).

In breaking down the world-wide estimate into regional impacts we have to remember that damages and casualties are not distributed in equal proportions. The death toll is far smaller in the developed world, which instead faces higher destruction damages. Rough country estimates by Smith (1992) suggest that the death toll per event is at least ten times lower in the developed world, compared to middle and low income countries.[21] This figure, together with the regional distribution of events calculated above, can be used to estimate the number of global warming-caused casualties. The results are shown in Table 3.14. For the monetary valuation, the value-of-life estimates derived earlier were used again. For the destruction estimate of column 3, the assumption on regional distribution was reversed, ie we presumed that the damage per event in the developed world is ten times the damage from an event in a developing country.

---

20 The estimates were compiled using the natural hazard map of the German reinsurance company Münchener Rück, reproduced in Berz (1990) and Smith (1992).

21 According to Smith the average number of deaths per event in Japan is 23, compared to 196 in middle income Philippines. In the extreme (and thus not necessarily representative) case of low income Bangladesh the average is 1,341.

Table 3.14 Costs from increased tropical storms

	Additional deaths	Value of lost lives (m$)	Destruction damage (m$)	Total costs (m$)
EU	0	0	0	0
USA	72	108	115	223
Ex-USSR	44	13	1	14
CHINA	779	78	13	91
OECD	313	470	506	976
WORLD	8,000	2,073	630	2,703

Source: see text.

TOTAL DAMAGE

In Table 3.15 the results of the previous section are summarized and added up to obtain the total damage for each region. With the exception of China and the former USSR total damage is in the order of about 1.5 per cent of GNP. About two-thirds of the total is non-market damage, ie damage which will not be reflected in the national accounts. Our estimates are slightly higher than those of Cline (1992a; for the US) and Nordhaus (1991b, c; US extended to the world), which both come up with a best guess of about 1 per cent of GNP, but are below Titus (1992), who estimates a 2xCO$_2$ damage of 2.5 per cent of GNP, again for the US. This latter estimate however assumes a 4°C temperature rise and is thus not directly comparable to our results. Tol (1993) estimates world-wide greenhouse damages at 1.9 per cent of GNP.

The figures are of course far from exact and one should allow for a range of error of probably at least ±50 per cent. We should also remember that several greenhouse impacts have not been quantified. These are probably predominantly harmful, with the possible exception of climate amenity. Overall, the results are thus clearly in the upper quarter of the Nordhaus range of 0.25 per cent to 2 per cent of GNP. *A more reasonable range is probably 1 per cent to 2 per cent of world GNP*, at least for developed countries and the world as a whole.

Greenhouse impacts are likely to be far more severe in developing countries, compared to the OECD. Table 3.16 provides a further dis-aggregation of the results. Leaving the special case of the former Soviet Union aside, our results predict a damage of about $70 bn in the non-OECD regions (China and Rest of the World). Although this is only about a quarter of total world-wide damage, it corresponds to about 2.2 per cent of GNP in these regions, some 60 per cent higher than the OECD average. The main causes for this high estimate are health impacts and the high portion of natural habitats and wetlands found in developing countries.

Table 3.15 Total damage due to 2xCO$_2$ (bn$)

	EU	USA	Ex-USSR	China	OECD	World
coastal defence	0.1	0.2	0.0	0.0	0.4	0.9
dryland loss	0.3	2.1	1.2	0.0	8.1	14.0
wetland loss	4.9	5.6	1.2	0.6	16.9	31.6
ecosystems loss	9.8	7.4	2.3	2.2	25.5	40.5
agriculture	9.7	7.4	6.2	7.8	23.1	39.1
forestry	0.1	0.6	0.4	0.0	1.8	2.0
fishery[a]	–	–	–	–	–	–
energy	7.0	6.9	–0.7	0.7	20.5	23.1
water	14.0	13.7	3.0	1.6	34.8	46.7
other sectors	–	–	–	–	–	–
amenity	–	–	–	–	–	–
life/morbidity[b]	13.2	10.0	2.3	2.9	34.4	49.2
air pollution	3.5	6.4	2.1	0.2	11.9	15.4
migration	1.0	0.5	0.2	0.6	2.0	4.3
nat. hazards[c]	0.0	0.2	0.0	0.1	1.0	2.7
TOTAL (bn$)	63.6	61.0	18.2	16.7	180.4	269.5
(% GNP)	(1.4)	(1.3)	(0.7)	(4.7)	(1.3)	(1.4)

a fishery loss is included in wetland loss
b mortality only
c hurricane damage only

Source: See text. Negative numbers denote benefits ('negative damage').

Table 3.16 2xCO$_2$ damage for different regions (present scale economy)

	bn $ (1988)	% GNP (1988)
European Union	63.6	1.4
United States	61.0	1.3
Other OECD	55.8	1.4
Former Soviet Union	18.2	0.7
China	16.7	4.7
Rest of the World	54.2	2.0
OECD	180.4	1.3
Non OECD	89.1	1.6
World	269.5	1.4

The situation could be further aggravated by a failure to implement the cost efficient adaption response (eg coastal protection), something which is quite likely to happen if the necessary funds are not made available. Although the data are weaker in the case of non-OECD countries, it seems therefore fair to say that global warming will have its worst impacts in the developing world, with a damage of *at least 2 per cent* of GNP for $2xCO_2$.

Regional differences can, however, be substantial, as is exemplified by the estimates for the former USSR and China. For the former Soviet Union damage could be as low as 0.7 per cent of GNP, about half the world average. Even this low level may come as a surprise to some people, however, as it has often been suggested that northern regions may benefit from global warming. Clearly, such a hope is fallacious. In the case of the former Soviet Union, possible positive impacts are more than offset by the costs of sea level rise, the particularly high health costs and the loss in consumer welfare due to higher world food prices. The extremely high estimate for China is caused by two factors, agricultural loss and life/morbidity impacts. Both of them are very volatile, and the probability range of total damage is therefore particularly wide for this country.[22]

CONCLUDING REMARKS

Chapter 3 has attempted to assess the damage associated with a doubling of atmospheric CO_2 concentration. Overall, the analysis confirms the damage range of 1 per cent to 2 per cent of GNP for $2xCO_2$ established in the previous chapter. The analysis is based on an enumerative approach, and the observations on this method, and damage estimation in general, made earlier are therefore also valid for the results presented here (see Chapter 2). Lack of data made it sometimes necessary to resort to some fairly *ad hoc*, and thus possibly subjective, assumptions. This has in particular affected the results for developing countries, and the confidence in these results is substantially lower than in those for developed regions.

On the whole, the 1 per cent to 2 per cent damage range for industrialized countries and the world as a whole is fairly robust, however, and the same appears to be true for the relative ranking of the different regions.

A word of caution is required with respect to the policy implications of the results. While the figures indicate a rather low damage with which at least the industrialized world should be able to cope, they do not necessarily imply that greenhouse policy action is unwarranted, for two reasons. First, the analysis has shown that, relative to their income, developing countries will be far more affected than their wealthy counterparts. Arguably, the destiny of the least well-off should be of particular concern to policy-makers, and the damages to developing countries should therefore be given additional weight

---

22 The picture is also less dramatic when damage is measured against purchasing power parity corrected GNP.

in the decision process.[23] Second, the estimates only concern one single point in time, ie shed light on the impacts of 2xCO$_2$ only. However, global warming will not stop there. 2xCO$_2$ could be reached as early as 2050, and what happens afterwards is as yet still unclear. Cline's work suggests that damage will increase exponentially with concentration (Cline, 1992a). Scientists speculate about the existence of discontinuities (see Chapter 2) and crossing certain ecological thresholds may well lead to unhappy surprises. Not least for these reasons, global warming still deserves our attention (see Chapter 9).

---

23 Note that the aggregate figures of world damage in this chapter and Chapter 4, do not make this adjustment. Somewhat arbitrarily, equal weight is given to all regions.

Chapter 4

The Marginal Social Costs per Unit Emitted

INTRODUCTION

Chapter 3 has dealt with the economic costs of greenhouse gas concentration doubling ($2xCO_2$). Despite the attention the $2xCO_2$ case enjoys in the literature, it is not directly relevant for practical purposes. For the appraisal of abatement projects it is more important to know the marginal costs of each additional tonne of emissions. Such an estimate is provided in this chapter.

Considerable effort has recently been put into analysing the social costs of the fuel cycle, with the aim of deriving externality adders which are to be put onto the price of fossil fuels to internalize the social costs of fuel consumption.[1] The studies typically concentrate on classic air pollutants like NO_x and SO_x. To complete the picture an additional adder would be required reflecting the social costs of global warming. A monetary estimate of the social costs of greenhouse gas emissions is also required to assess individual greenhouse gas abatement projects such as those financed by the Global Environment Facility (GEF).

The aim of this chapter is to fill this gap and provide an order of magnitude assessment of the social costs of greenhouse gas emissions. Assessing greenhouse damage is not possible without accounting, in one way or another, for the huge uncertainty prevailing in the global warming debate. Although scientists have achieved a remarkable consensus with respect to many aspects, our ignorance of global warming impacts is still vast, particularly with respect to regional and long-term impacts. Most studies allow for uncertainty by working with different climate scenarios. In this chapter we choose a different approach and incorporate uncertainty directly by describing uncertain parameters as random.

Using a stochastic model has several advantages. First of all it allows a better representation of current scientific understanding. Scientific predictions usually take the form of a best guess value supplemented by a range of possible outcomes. Concentrating on the best guess value therefore neglects a large part of the information provided, while, on the other hand, a stochastic model can make full use of it. Secondly, and probably more

---

1 See eg Hohmeyer (1988), PACE (1990), Pearce et al. (1992), and Lockwood (1992).

importantly, a stochastic model allows the calculation of an entire damage probability distribution, thereby providing important additional information on the likelihood of the estimates and the possibility of extremely adverse events.

Care should nevertheless be exercised when interpreting the figures presented below. Although, as we believe, based on the best available scientific information, they cannot provide anything better than a rough order of magnitude assessment.

PREVIOUS SOCIAL COST ESTIMATES

Actual social costs vs shadow values

Before discussing previous studies on the social costs of greenhouse gas emissions, it is helpful to draw a distinction between figures on the *actual marginal social costs* of greenhouse gas emissions and the concept of a *shadow price*. Most studies estimating the social costs of greenhouse gas emissions do so in an intertemporal optimization framework. That is, their primary concern is to calculate socially optimal emission levels at each point in time. In such a set-up the shadow price of emissions is defined as the pollution tax required to keep emissions at the optimal level. For certain types of models this value is related to the social costs of emissions.

As will be discussed in Chapter 9, there are two main economic approaches to the question of the optimal global warming policy response, the cost-benefit approach and the carbon budget approach.

The cost-benefit approach (CBA)

In a cost-benefit framework the optimal outcome is obtained at the inter-section point of the marginal cost and the marginal benefit curves. That is, in the case of global warming, the incremental costs of additional greenhouse gas abatement have to be equal to the additional benefits of avoided damage, and this at each point in time. This situation is achieved by taxing emissions at a level equal to the marginal global warming damage they cause. The shadow price of emissions is thus equal to the actual social costs. However, since marginal damage depends on the amount of emissions discharged in the future, this is strictly speaking only the case if future emissions follow the optimal emission trajectory calculated in the model. There is no guarantee that this will be the case. If future emissions deviate from the optimal path, shadow values and actual social costs will differ. The discrepancy should only be slight, however, and the shadow values calculated in cost-benefit models can serve as a good indicator of the actual social costs of greenhouse gas emissions.

The carbon budget approach

An alternative approach is to ignore damage considerations and exogenously impose an upper atmospheric concentration or warming limit, determined on the basis of ethical, political or precautionary considerations. Under this method the shadow value of emissions will reflect the costs of the additionally imposed constraint, and will have no connection to the actual greenhouse damage occurring. This has the advantage that climate and greenhouse damage need not actually be modelled. On the other hand the question remains open as to how stringent the carbon constraint should be. The shadow values from carbon budget models are often significantly higher than actual social cost estimates, at least in later time periods. Anderson and Williams (1993), for example, propose a carbon tax starting at $25/tC in 1990 and rising to $120/tC by 2010. The reason for this high tax is that carbon budget studies usually assume a rather stringent carbon constraint. In the case of Anderson and Williams (1993), for example, the aim is to introduce an economical carbon-free energy source by the year 2010.

Previous results

The shadow value of cost-benefit models can thus be used as an indication of the social costs of greenhouse gas emissions. Alternatively, the social costs can be calculated directly as the difference in future damage levels caused by a marginal change in baseline emissions. Both methods have been used so far.

The pioneering paper on the social costs of CO_2 emissions is Nordhaus (1991b, c). Using a simplified version of a dynamic optimization model, he calculates social costs of $7.3 per tonne of carbon emitted. Imposing different assumptions on the rate of discount and the $2xCO_2$ damage leads to a range of $0.3/tC to $65.9/tC. Implying that abatement should only be undertaken as long as costs do not exceed $7.3 per tonne of carbon abated, the estimates formed the backbone of Nordhaus' claim that global warming may not, after all, be such a big problem, and may justify only a modest policy response (see Chapter 9).

The Nordhaus estimates have been strongly criticised by several authors, including for example Ayres and Walter (1991), Daily et al (1991), Cline (1992a) and Grubb (1993). While the main objection concerned Nordhaus' $2xCO_2$ damage estimate, the more substantial shortcomings are probably those concerning the model itself (see Cline, 1992a). Especially controversial is the assumption of a *resource steady state*, which *inter alia* implies a constant level of CO_2 emissions over time. Obviously this is unrealistic. The IPCC, for example, predicts an increase in annual CO_2 emissions from about 7 GtC in 1990 to about 9–14 GtC by 2025 (IPCC, 1992a). The simple (linear) structure of the climate and damage sectors also implies that costs will remain constant at $7.3/tC throughout. Climate processes are clearly non-linear, and the costs of CO_2 emissions will thus depend on future concentration and warming levels, ie they will vary over time. Subsequent estimates suggest that they may in fact rise over time. That is, a tonne of CO_2 added to an already large stock of atmospheric CO_2 is likely to cause a higher damage than a tonne emitted under a low concentration level.

These objections are also relevant to the study by Ayres and Walter (1991), whose calculations are based on the Nordhaus model. The paper has additional shortcomings. Particularly questionable is the fact that the authors use identical commodity values, eg for land, in all countries of the world. This is unrealistic. Land prices in Europe clearly differ from those, say, in India or Pakistan. By considering both the costs of sea level rise protection and the costs of climate refugees from coastal regions, Ayres and Walter further appear to double count at least some of the sea level rise impacts. On the whole, their cost estimate of $30–$35/tC is therefore doubtful.

The shortcomings of the earlier model were recognized and corrected in Nordhaus' subsequent approach, the DICE (Dynamic Integrated Climate Economy) model (Nordhaus, 1992, 1993a, b). DICE is an optimal growth model in the Ramsey tradition, extended to include a climate module and a damage sector which feed climate changes back to the economy. The shadow values of carbon following from DICE are in the same order as Nordhaus' previous results, starting at $5.3/tC in 1995 and gradually rising to $6.8/tC in 2005 and $10/tC in 2025 (see Table 4.1). Note that figures for future periods are current value estimates, ie they denote the social costs valued at the time of emission.

The DICE model was also used by Cline (1992b), who concludes that Nordhaus' choice of parameter values may have led to an underestimation of

Table 4.1 Existing estimates of CO_2 emission costs ($/tC)

Study	Type	1991–2000	2001–10	2011–20	2021–30
Nordhaus (1991b, c)	MC		7.3 (0.3–65.9)		
Ayres and Walter (1991)	MC		30–35		
Nordhaus (1993a, b)	CBA	5.3	6.8	8.6	10.0
Cline (1992b, 1993a)	CBA	5.8–124	7.6–154	9.8–186	11.8–221
Peck and Teisberg (1992)	CBA	10–12	12–14	14–18	18–22
Maddison (1993)	CBA/MC	5.9–6.1	8.1–8.4	11.1–11.5	14.7–15.2

Note: MC = actual marginal costs,
CBA = shadow value of a cost-benefit study.

the true costs. Of particular importance is the discount rate. The shadow values of the Cline replication are reported in Cline (1993a), and are also shown in Table 4.1.

Figures within the same order of magnitude as the original DICE results were suggested by Peck and Teisberg (1992, 1993a, b) and Maddison (1993). Peck and Teisberg came up with a shadow value of carbon of about $10/tC in 1990, rising to about $22/tC by 2030. The CETA (Carbon Emission Trajectory Assessment) model, on which their calculations are based, possesses a similar climate and damage sector as DICE, but is more detailed on the economy side by incorporating a carefully modelled energy sector.[2] Differences between the estimates appear to be mainly due to different assumptions about the extent of $2xCO_2$ damage. Common to all three papers is the assumption of a 3 per cent utility discount rate. The results are highly sensitive to this assumption, as Cline's results have underlined.

THE SOCIAL COSTS OF GREENHOUSE GAS EMISSIONS

Model description

In the remainder of this chapter we will introduce our own estimates of the social costs of greenhouse gas emissions, based on Fankhauser (1994a). They are actual marginal cost figures, rather than shadow values. The difference is explained in Figure 4.1. In addition to CO_2, the costs of methane and nitrous oxide will also be considered. The basic features of the model are as follows (see also Appendix 1).

Greenhouse gases are so-called stock pollutants. That is, global warming damage is not caused by the flow of emissions as such, but by their accumulation in the atmosphere. Consequently a tonne of emissions has its impact not only in the period of emission, but over several time periods – as long as the gas, or fractions of it, remains in the atmosphere. The damage costs of a tonne of emissions are thus a present value figure – the discounted sum of additional future damages which occur as a consequence of the emission increase.

Contrary to optimization models, future emissions are exogenously determined in this model. The social costs of emissions in a certain period are achieved by increasing emissions by one tonne in this period and comparing the stream of annual damages before and after the increase.

A model to estimate such a figure requires several elements. First we require a module which represents the world climate system and transforms an increase in emissions into incremental atmospheric concentration and then warming. The climate system being a highly complex process, models aiming at its simulation, such as Global Circulation Models (GCMs) tend to

---

2 This latter feature is of course of less relevance in the present context.

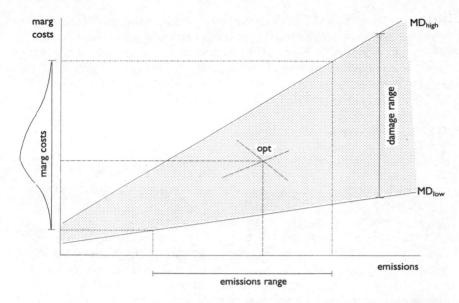

Marginal damage costs depend on the level of emissions. In the usual case of an upwards sloping MD curve they are rising with the level of emissions, as shown in the Figure. Optimization models calculate marginal costs at the point where marginal damage equals marginal abatement costs – the dashed situation labelled 'opt'. The present model calculates marginal costs at the emission level actually observed. Since emissions are uncertain, and so is the exact shape of the marginal damage curve, calculations will result in a *range* of possible cost figures, each of which is assigned a certain probability of occurrence – the probability distribution shown to the left of the figure.

Source: Fankhauser (1994a).

Figure 4.1   Actual marginal damage vs shadow values

be quite big. Smaller-scale representations are often able to provide a sufficiently close approximation, though, and our model is based on such an approximation (see Appendix 1).

Second, we need to know more about the shape of the damage function. Research has concentrated almost solely on $2xCO_2$, and information about the damage before and after this benchmark is therefore extremely difficult to obtain. Assumptions are often rather *ad hoc*. Long-run damage in particular also depends heavily on the assumed discount rate, the choice of which is a highly controversial issue (see Chapter 8). Finally, because the climate process is non-linear, the marginal costs depend on the level of emissions in future periods, and we need predictions about these.

Future emissions, discount rate, shape of damage function and many of the climatic parameters are highly controversial variables about which little is known with certainty. Uncertainty is explicitly taken into account in our model by modelling all key parameters as random. In the basic case we assumed triangular distributions for all variables. The triangular specification appeared to be an obvious choice, given that most scientific predictions take the form of a lower bound, upper bound and best guess estimate, values which can directly be used as distribution parameters without requiring further alterations. Alternative distributions will later be introduced for some parameters to incorporate low probability/high impact events. A more detailed description of the model is provided in Appendix 1.

The social costs of CO_2 emissions

We have used the model for simulations of the social costs of CO_2 emissions over four decades – from 1991 to 2030. The results are shown in Table 4.2. As expected, damage per tonne of emissions is rising over time, from about $20/tC between 1991 and 2000 to about $28/tC in the decade 2021–30. The rise is mainly due to income and population growth, which causes absolute damage levels to rise over time. The impact of higher future concentration levels, on the other hand, is ambiguous. In some constellations the logarithmic relationship between radiative forcing and concentration dominates the result, and a higher concentration leads to a decrease in marginal damage. If it was not for economic and population growth, marginal costs

Table 4.2 The social costs of greenhouse gas emissions

	1991–2000	*2001–10*	*2011–20*	*2021–30*
CO_2				
mean ($/tC)	20.3	22.8	25.3	27.8
5th percentile	6.2	7.4	8.3	9.2
95th percentile	45.2	52.9	58.4	64.2
standard dev.	14.3	16.0	17.5	19.0
skewedness	2.5	2.5	2.5	2.4
CH_4				
mean ($/tCH$_4$)	108	129	152	176
5th percentile	48	58	69	79
95th percentile	205	249	293	342
standard dev.	54	64	75	89
skewedness	1.6	1.6	1.7	1.8
N_2O				
mean ($/tN)	2,895	3,379	3,901	4,489
5th percentile	805	953	1,122	1,277
95th percentile	7,253	8,361	9,681	11,121
standard dev.	2,895	2,595	2,996	3,455
skewedness	2.2	2.2	2.3	2.3

Source: Fankhauser (1994a).

would fall over time in these cases. The figures for future periods are again current value estimates and denote the social costs valued at the time of emission.

The expected value figures alone do not of course tell a complete story. The optimal policy response is likely to differ depending on the confidence in the results, the distribution of possible outcomes and the probability of high impact events. What is lacking is thus some information about the probability distribution of greenhouse damage. Such a probability distribution can be obtained directly from our stochastic model, and the relevant statistics are also shown in Table 4.2. The distributions for CO_2 emissions in two different decades are depicted in Figure 4.2.

The figure shows rather wide distributions with standard errors around 14 to 19, reflecting the generally low level of confidence in these figures. Not surprisingly, the standard error is increasing over time, as the estimates for more distant periods are less certain than that for the decade 1991–2000. The shape is clearly asymmetric and skewed to the right, with coefficients of skewedness in the order of 2.5 (see Table 4.2).[3] Loosely, this means that the probability of an extremely disastrous outcome is higher than that of an extremely modest result.

Our damage estimates are somewhat higher than those of existing studies like Nordhaus (1992, 1993a, b) and Peck and Teisberg (1992, 1993a, b) (see Table 4.1). Partly this is due to different assumptions on the value of some key parameters. The utility discount rate, for example, is set at 3 per cent in DICE and CETA, a value which constitutes the upper bound for this parameter here. On the other hand, we used more moderate assumptions about the slope of the damage function.

Conceptually more important is a second source of discrepancy, which arises from the fact that our figures represent *expected values*, while the other estimates are *best guesses*. As shown in Figure 4.2, global warming damage is not distributed symmetrically but skewed to the right. Under these circumstances the mean will be greater than the mode, and expected value figures are therefore bound to be higher than a best guess estimate which ignores this asymmetry. The higher value of our figures is thus also a consequence of the incorporation of high impact events. In our model the difference between the expected value and a non-random best guess is about 25 per cent.

Other greenhouse gases

CO_2 is not the only greenhouse gas, despite being by far the most important one, accounting for more than half of the total effect. Other gases contributing to the greenhouse effect include CH_4, N_2O and CFCs. The usual way of dealing with greenhouse gases other than CO_2 is to transform them

---

3   For comparison, the skewedness of a symmetric distribution is zero.

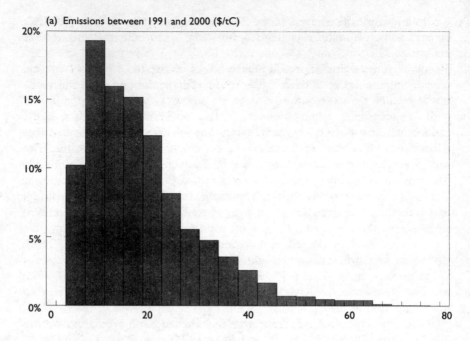

(a) Emissions between 1991 and 2000 ($/tC)

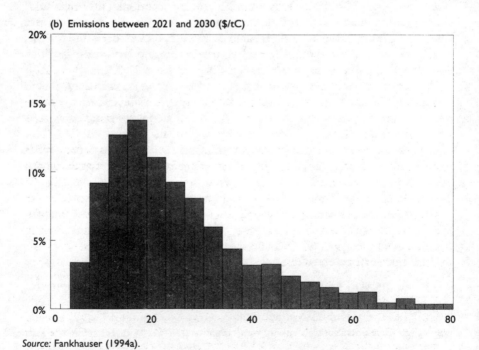

(b) Emissions between 2021 and 2030 ($/tC)

Source: Fankhauser (1994a).

Figure 4.2 The probability distribution of CO_2 damage

into CO_2-equivalents by using Global Warming Potentials (GWPs). According to this index, suggested by the IPCC, one tonne of CH_4 and N_2O are equivalent to 11 and 270 tonnes of CO_2, respectively (IPCC, 1992a). However, it has repeatedly been pointed out in the literature that GWPs are not an adequate index if we are concerned with the damage caused by each gas.[4]

GWPs are a measure of the relative radiative forcing capability of gases. Relative forcing, however, corresponds to the relative damage potential only if the latter is a linear function of the former. This is not the case. The incremental damage caused by a marginal increase in radiative forcing depends on factors like previous levels of radiative forcing and the degree of warming already encountered. The relative damage contribution of a gas will therefore vary over time and in general be different from its relative forcing capacity.

To exemplify this point, we have computed the social costs of other greenhouse gases directly by calculating the present value marginal damage caused by a marginal increase in other greenhouse gas emissions. Table 4.2 shows the results for methane and nitrous oxide. Again the figures are rising over time, in the case of methane from $108/tCH_4 in 1991–2000 to $176/tCH_4 in 2021–30. The damage from nitrous oxide rises from $2,895/tN in 1991–2000 to $4,489/tN.

In terms of CO_2 equivalents or 'Greenhouse Damage Potentials' this corresponds to values of 20–23 for CH_4 and 333–377 for N_2O, depending on the time of emission.[5] They are roughly in line with the estimates of Hoel and Isaksen (1993), but, for the reasons given above, are considerably different from GWPs. The fact that they are higher than GWPs mainly reflects the impact of discounting, which lessens the long-run impacts of long-lived gases like CO_2. The figures are restricted to the direct impacts of a gas and neglect indirect effects such as the impact of methane on water vapour and tropospheric ozone.

Indirect effects would be particularly important for CFCs, for which it has been estimated that as much as 80 per cent of the initial contribution may be offset through the indirect cooling effect from ozone depletion (IPCC, 1992a; Wigley and Raper, 1992). This recent finding has considerably downgraded the importance of CFCs as greenhouse gases. For this reason, and because their phasing out has by and large already been agreed upon in the Montreal protocol, CFCs are not considered here.[6] Also note that the figures only refer to the social costs from global warming, ie they do not

---

4 Reilly (1992), Schmalensee (1993), Hoel and Isaksen (1993).
5 As with Global Warming Potentials, the figures compare the impact of one tonne of each gas, ie they compare a tonne of CH_4 or N_2O with a tonne of CO_2. The figures in Table 4.2 on the other hand use the more common measures of tCH_4 for methane, tN for N_2O, and tC for CO_2.
6 Figures for CFCs can be found in Hoel and Isaksen (1993).

include costs from other environmental problems. Again these would be most important in the case of CFCs.

SENSITIVITY ANALYSIS

Discounting

One of the key parameters in the above estimates is the discount rate, not only because of the ongoing controversy about its appropriate level, but also because of the high sensitivity of the results with respect to it. We will discuss discounting in more detail in Chapter 8. The model here uses the consumption equivalent technique, as described in, for example, Lind (1982). It involves the transformation of all damage into consumption equivalents, and then discounting at the consumption rate of discount. See Appendix 1 for details.

The consumption rate of discount consists of two parts. The first accounts for the effect of decreasing marginal utility of income. As individuals become richer over time, they will gain less and less additional satisfaction from an additional dollar of income. The second element of the discount rate represents people's impatience or myopia: consumption today is considered superior to consumption tomorrow.

The main controversy in the discounting debate is about this second element, called the pure rate of time preference or utility discount rate. Loosely, there are two schools of thought (see Chapter 8). The first argues in favour of a zero rate, on the grounds that a positive rate would unjustifiably attach less importance to future generations. The second school points at empirical evidence in favour of a positive rate. The sensitivity analysis concerns this parameter, labelled ρ.

The results in Table 4.2 were obtained assuming a triangular probability distribution for the pure rate of time preference with upper and lower bounds of 0 per cent and 3 per cent, respectively, and a best guess value of 0.5 per cent. To illustrate the impacts of different discount rates we have re-run the model for ρ held fixed at different values. The results are summarized in Table 4.3.

If ρ is set at 3 per cent the expected shadow value of 1990 CO_2 emissions falls from \$20/tC to a mere \$5.5/tC. By 2030 the figure has risen only slightly to \$8.3/tC. Interestingly, this is almost exactly the Nordhaus (1992; 1993a, b) result. Assuming a pure rate of time preference of 0, on the other hand, yields an expected value of \$48.8/tC in 1990, and \$62.9/tC in 2030, about eight times more than under the high rate. Using a higher rate also leads to a considerable reduction in the standard deviation, and thus the confidence interval. This is because more uncertain future impacts are weighted less under a higher discount rate. The lower skewedness, on the other hand, is mainly due to the fact that parameter ρ, which had a skewed distribution in the random case, is now fixed.

Table 4.3   The social costs of CO_2 emissions – the impact of discounting

		1991–2000	*2021–30*
Random Case	mean ($/tC)	20.3	27.8
-> ρ = (0,0.005,0.03)	5th percentile	6.2	9.2
	95th percentile	45.2	64.2
	standard dev.	14.3	19.0
	skewedness	2.5	2.4
Low discounting	mean ($/tC)	48.8	62.9
-> ρ = 0	5th percentile	27.6	34.9
	95th percentile	80.1	104.6
	standard dev.	15.6	22.4
	skewedness	0.9	1.3
High discounting	mean ($/tC)	5.5	8.3
-> ρ = 0.03	5th percentile	3.7	5.3
	95th percentile	7.6	12.0
	standard dev.	1.2	2.1
	skewedness	0.5	0.8

Source: Fankhauser (1994a).

The high sensitivity of the results with respect to discounting should come as no surprise. It is a direct consequence of the long-term character of global warming and the fact that damages will only occur several decades into the future. The results clearly underline the importance of the discounting question, though, and the crucial role that ethical issues ought to play in the future debate on global warming.

Catastrophic events

Although the parameter values underlying the above results broadly reflect the current understanding of global warming, there is still an element of subjectivity inherent in them. In particular, by assuming a triangular distribution for random parameters they neglect the possibility of a climate catastrophe. It has often been noted that, given the complexity of the climatic system and the unprecedented stress imposed on it, a catastrophic result cannot be excluded with certainty, particularly in the long run (beyond $2xCO_2$). Worst case scenarios implied in the literature include the melting of the antarctic ice-sheet, a redirection of the gulf stream and the release of methane from previously frozen materials through the melting of permafrost soils (see Chapter 2). The probability of a catastrophic outcome is clearly greater than zero.

The easiest way to incorporate such low probability/high impact events is by using probability distributions with a domain greater than zero, ie to assume that parameter values are bounded from below but not from above. Even extremely high parameter values then still occur with a positive probability. A distribution with this property is the lognormal, and as a sensitivity test we have run the model assuming a lognormal distribution for three key parameters: climate sensitivity, $2xCO_2$ damage and the slope of the damage function, thus allowing for catastrophic outcomes with respect to climate, with respect to impacts and with respect to the existence of a threshold.[7] The distributions were calibrated such that the lower bound remains unchanged and the most likely value equals the scientific best guess, as before, while the probability of an extremely unfavourable outcome is gradually increased. The probability of the initial upper bound being exceeded rises from practically zero in the most optimistic case (scenario A) to 5 per cent, 10 per cent and then 15 per cent.[8] The results of this exercise are summarized in Figure 4.3.

Figure 4.3 shows the mean and 90 per cent confidence interval of the marginal social costs of CO_2 emissions between 1991 and 2000, under the different scenarios considered. Moving from scenario A, which roughly corresponds to the triangular case used before, to scenario D, in which there is a 15 per cent chance of the initial upper bound values being exceeded, increases the expected social costs by about 60 per cent, from $20/tC to $33/tC. As expected, the 95th percentile rises stronger than the mean, by about 80 per cent, thus further increasing the skewedness of the distribution. Although illustrative, the analysis therefore clearly underlines the importance of low probability/high impact events.

POLICY IMPLICATIONS

This chapter provides estimates of the monetary costs of greenhouse gas emissions. To recapitulate: as a rough benchmark figure we suggest a value of $20/tC for emissions between 1991 and 2000. In subsequent decades the value rises to $23/tC, $25/tC and finally $28/tC for emissions in the third decade of the next century. Like all greenhouse damage estimates these results are highly uncertain and the confidence intervals attached to them are correspondingly wide. The stochastic character of our model allowed the explicit calculation of a damage probability distribution. It was shown that the distribution is skewed to the right, even for the runs neglecting the possibility of a climate catastrophe. That is, even when abstracting from actual extremes, a disastrous outcome is still more likely to occur than a

---

7 An extremely steep damage function can be seen as an approximation of a threshold at $2xCO_2$, see Appendix 1.
8 The upper bound specifications in the triangular case were 4.5°C for the climate sensitivity parameter, 2 per cent of GNP for $2xCO_2$ damage and a power of 2 for the damage function.

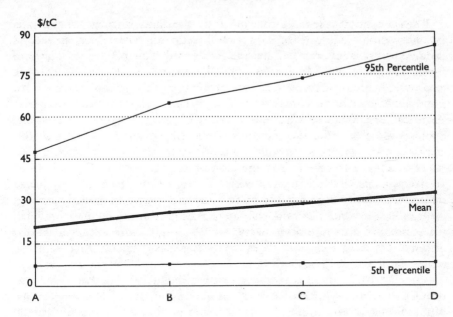

Moving from scenario A to scenario D, the likelihood of the initially specified upper bound values being exceeded rises from 0 per cent to 5 per cent, 10 per cent and 15 per cent. An alternative interpretation is as follows: in each scenario and for each of the three parameters considered, there is a 1 per cent probability that the actual value will exceed the figures given in the table below. For example, in scenario B there is a 1 per cent probability that $2xCO_2$ damage is greater than 3 per cent of GNP.

	Climate sensitivity	*$2xCO_2$ damage*	*Power of damage function*
Scenario A	3.5°C	1.7% GNP	1.5
Scenario B	5.5°C	3.00% GNP	2.5
Scenario C	6.0°C	3.50% GNP	3.0
Scenario D	7.0°C	4.25% GNP	3.5

Source: Fankhauser (1994a).

Figure 4.3 CO_2 damage distribution and the threat of catastrophe ($/tC, 1991–2000 emissions)

correspondingly modest result. Incorporating the possibility of a future climate catastrophe considerably increases both the mean and the skewedness of the distribution. In the most extreme case considered expected damage rose to about $33/tC. It was also confirmed that the results crucially depend on the choice of the discount rate, and ethical considerations will therefore have to stage prominently in the future debate.

The main application for the estimates is probably project appraisal. For small projects the interpretation of the figures is straightforward. For a reforestation project sequestering 1 mtC per year over 30 years, for example, we can expect benefits of $20 m/yr in the first decade, $23 m/yr in the second and $25 m/yr in the third. Total (undiscounted) benefits are therefore (in million dollars) $200 + 230 + 250 = 680$. Investment decisions can then be made in the usual way by comparing the relative net benefits of rival projects. The analysis is more complicated with respect to large-scale abatement policies big enough to affect the future emission trajectory. Because the shadow value of carbon depends on future emissions the social costs of CO_2 emissions will change with the implementation of the policy and would have to be re-calculated for the new emission trajectory.

The appraisal of individual abatement projects has to be distinguished from the task of designing an optimal policy response to global warming. Our model does not deal with this latter question, and the figures provided therefore give little indication of the socially optimal carbon tax. A socially optimal emission trajectory is, on the other hand, calculated in optimal control model like CETA or DICE.[9] Both models provide a first assessment as to what the optimal emission trajectory might be. They are however, limited to the hypothetical case of perfect information, and neglect the implications of uncertainty. We will come back to the issue of optimal carbon abatement in Chapter 9.

---

9 See Pack and Teisberg (1992, 1993a, b), Nordhaus (1992, 1993a, b) and Cline (1992b).

Chapter 5

The Costs of Adaptation: The Case of Sea Level Rise

INTRODUCTION

Chapter 5 analyses the damage costs from a climate-induced increase in sea level in more detail. The case of sea level rise (SLR) is of interest for at least two reasons. First, SLR is likely to be one of the most crucial and harmful impacts of global warming, as shown in Chapter 3. In the United States, for example, almost 50 per cent of the population lives within 24 kilometres of the coast (Turner et al, 1994b). Second, SLR well illustrates the fact that parts of greenhouse costs will not actually be damage costs as such, but will arise from the implementation of damage mitigation (or adaptation) strategies, such as the erection or modification of sea defences.

The costs of global warming will crucially depend on the extent to which measures are undertaken to accommodate to global warming (see Chapter 2). A first concern of this chapter is therefore the question of optimal adaptation. In general, adaptation should take place as long as the benefits from avoided damage exceed the incremental costs of additional action. In the case of SLR the trade-off is between the costs of SLR protection and the benefits from land loss avoided, but it should also take into account that protection walls lead to a reduction in the damage from storm surges, while on the other hand they add to the already existing pressure on the natural environment.

Most damage studies do not explicitly model this trade-off and instead assume an exogenously given rule about the optimal level of protection, usually a partial retreat scenario in which developed areas are protected, while sparsely populated or low value land is abandoned (see eg IPCC, 1990c). While this seems a reasonable rule of thumb, there is no guarantee that it will hold in general. The optimal degree of protection may differ between regions as well as for different degrees of SLR. For example, we may find lower optimal protection levels in poorer countries, where the costs of protection may be relatively high compared to the value of land at threat.[1] At first sight we may also expect that a high SLR will lead to a higher degree

---

1 It is often, and correctly, pointed out that poor countries may lack the funding to implement sufficient protection measures and, as a consequence, will be underprotected (see Chapter 3). This is not the point made here. Here we argue that even if funds were available it may not be efficient to achieve the same degree of protection as in a rich country.

of protection, since a larger part of the hinterland is now at threat and the potential land loss damage is higher. But then, protection costs per kilometre of coast will also increase, since higher protection measures are now required, and this would point towards a lower degree of protection. The net effect is therefore unclear. To shed further light on such aspects, the optimal degree of protection is determined endogenously in this chapter.

The second concern of the chapter is with an overall assessment of SLR damage. Designing the optimal policy response to SLR is mainly a problem of regional coastal zone management, and the optimal protection strategy will generally be different for different coastlines. Ideally the global picture would thus emerge from the aggregation of the many existing local assessments.[2] However, it is evidently not feasible to study all vulnerable coastlines in the required detail, and a global assessment of SLR damage will therefore necessarily have to be based on a 'top down' approximation.[3] Several attempts at such a 'top down' assessment have already been made.[4] These studies typically concentrate on just one or two scenarios, eg on the benchmark case of a 1 metre rise by the year 2100. The analysis here goes further in that it assesses damage caused by different levels of SLR. The chapter also explicitly pays attention to the gradual character of SLR, which will occur slowly over time.

The next section provides a general outline of the role of damage, adaptation and abatement costs within a cost-benefit framework. The special case of SLR is then analysed in the subsequent sections.

GREENHOUSE GAS ABATEMENT VS. ADAPTATION: THE GENERAL CONTEXT

Loosely, policy-makers have two sets of options to moderate the impacts of greenhouse warming. They can either limit the (net) amount of greenhouse gases emitted – the well-known abatement and sequestration options, leading to a lower degree of warming – or, they can ease the impacts of a given change through appropriate protection/adaptation measures. Although we concentrate on the example of SLR here, adaptation is in no way limited to SLR protection, but may play an important role for other damage aspects as well, as discussed in Chapter 2. Both sets of options – abatement/sequestration and protection/adaptation – have to be taken into account when drafting the greenhouse policy response.

Analytically, the optimal combination of abatement and adaptation can be found by minimizing the total costs of climate change, consisting of the costs

---

2 See eg Turner et al (1994a) on East Anglia; Gleick and Maurer (1990) on the San Francisco Bay; Milliman et al (1989) on the Nile and Bengal deltas; den Elzen and Rotmans (1992) and Rijkswaterstaat (1991) on the Netherlands; and IPCC (1992b) on several regions.

3 The ongoing attempts to derive a global picture from local case studies are summarized in IPCC (1992b).

4 Eg IPCC (1990c, 1992b), Rijsberman (1991), and Titus et al (1991).

of emission abatement (or sequestration) AC,[5] the costs of protection measures P and the costs of greenhouse damage D, ie

$$\min_{m,e} \ AC(e) + P(m) + D(T,m) \tag{1}$$

subject to

$$T = f(e) \tag{2}$$

where e denotes the level of abatement, and m the degree of adaptation. Climate change is symbolized by the variable T (for temperature change), and depends negatively on the amount of abatement e. The more abatement is carried out, the lower will be the temperature increase. Greenhouse damage D depends positively on the temperature level, and negatively on the amount of adaptation, m. AC and PC both depend positively on their arguments.

The optimal conditions for this simultaneous optimization problem require that the marginal costs of each policy measure be equal to the benefits of abatement and adaptation, respectively. This is the standard cost-benefit result. *Further abatement efforts are justified as long as the incremental costs of doing so are smaller than the additional benefits from warming avoided. Similarly, further adaptation is warranted as long as the additional costs are lower than the additional benefits from mitigated damages.* Graphically, we seek for both policy measures the intersection of the respective marginal cost and benefit curves (see Figure 5.1).

The point to note is that the optimal value for each of the options, e^* and m^*, depends on the value chosen for the other. For example, if significant abatement measures are taken, little action may be needed with respect to adaptation. Conversely, if the consequences of global warming can easily and cheaply be accommodated by adaptation, there may be less need for preventive carbon abatement. The two sets of policy measures should thus be carefully coordinated, and would ideally be determined simultaneously.

In the real world this is hardly ever the case. Decisions about e^* and m^* are usually taken at different political levels. The question of optimal emissions is in general addressed globally, in international negotiations between countries or at international conferences such as UNCED 1992. The composition of the optimal adaptation strategy, on the other hand, is often left to local authorities – in the case of coastal protection, for example, to the regional coastal zone managers. The same division is also reflected in the literature, where studies typically either deal with the optimal abatement question[6] or with the optimal level of adaptation, usually in the context of sea level rise protection.[7]

---

5 Climate change prevention policies may also include geo-engineering solutions such as seeding of the oceans to induce carbon uptake. We abstract from these possibilities here.
6 Nordhaus (1991b, c, 1992, 1993a, b), Peck and Teisberg (1992, 1993a, b), Cline (1992a, b).
7 Eg Turner et al (1994a), Gleick and Maurer (1990).

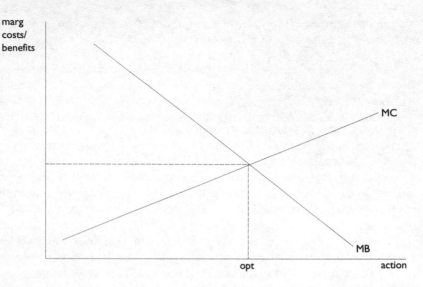

For both types of global warming policies, abatement and adaptation, further action is warranted as long as the marginal costs of the respective action, MC, are lower than the marginal benefits MB. For each policy, the optimum is reached at the point where the two curves intersect.

Figure 5.1   The optimal level of greenhouse action

The following analysis will show that this two-step process of decision-making will lead to the same result as a simultaneous treatment, provided that the damage function in the second, global, step is appropriately defined. To analyse the local problem of optimal adaptation then, it is easiest to interpret regions as *climate takers*. The greenhouse gas emissions of an individual region or country are in most cases so small that they have virtually no influence on world climate (Adger and Fankhauser, 1993). Climate, and climate change, is therefore exogenously given, and the regional decision problem is merely concerned with finding the optimal degree of protection against this change,

$$\min_m \quad P(m) + D(m \mid T = f(e)) \tag{3}$$

Again the first order condition will require that marginal protection costs equal the marginal benefit from damage avoided. Because D is a function of the exogenous climate change (and thus ultimately of e), the optimal adaptation level, m^*, will also be a function of T: $m^*(T)$.

As a next step we substitute m^* back into (3) to obtain a function called V. V denotes the combined adaptation and damage costs of climate change caused by a level of change T, *given that the optimal policy response has been taken with respect to adaptation*. In formal notation:

$$V(T) = P[m^*(T)] + D[m^*(T), T] \qquad (4)$$
$$= \min_m[P(m) + D(m; T)]$$

Let us now turn to the global problem of optimal abatement. Intuitively it should be clear that it is this function V, or its aggregation over all regions, which is relevant to the global problem and that the correct way of deciding the optimal abatement level is to trade off the costs of abatement AC with the minimal combined damage costs V. That is, the global problem is:

$$\min_e \quad AC(e) + V(T) \qquad (5)$$

where again T=f(e). It is easily shown (Fankhauser, 1994b) that problem (5) has the same solution as problem (1). The marginal costs of each policy measure have to equal marginal benefits.

The two-step process in which decisions on adaptation are taken locally, while the optimal abatement level is determined at a global level, is thus equivalent to a simultaneous optimization, *provided that the global warming damage relationship is specified as in equation (4), as the cost minimizing combination of adaptation plus damage costs.*

Most damage studies acknowledge this property or at least pay lip service to it. However, when it comes to actual damage calculations, data limitations do not always allow the careful implementation of this principle (see Chapters 2 and 3). The resulting estimates will then overestimate the true damage. This was illustrated in Chapter 2 for the case of agriculture (see Table 2.2). However, the definition of the damage function in (4) also points out potential sources for an underestimation. The picture is only complete if both damage and protection costs are considered, and for many studies this is not the case. The assessment of human amenity impacts in Chapter 3, for example, was limited to the defence costs, which occur in the form of a higher electricity demand from air cooling. We were unable to capture the remaining amenity impacts of warming.

In the case of SLR, as mentioned above, the assumption usually is that m^* corresponds to a partial retreat strategy. In the remainder of this chapter we attempt to derive a minimum function like V(T) for SLR. To do this we first analyse the local problem of optimal protection for the case of SLR.

OPTIMAL SEA LEVEL RISE PROTECTION

Problem description and model

Finding the optimal SLR protection strategy is not as easy as may have been suggested in the previous section. Instead of the single and continuous variable m, decision-makers will be confronted with a host of alternative strategies. In its 1990 report the coastal zone management subgroup of the IPCC divided available SLR response options into three groups, with the following definitions (IPCC, 1992b, p 8; see also IPCC, 1990c):

Retreat – abandon structures in currently developed areas, resettle the inhabitants, and require that any new development be set back specific distances from the shore, as appropriate.

Accommodate – continue to occupy vulnerable areas, but accept the greater degree of flooding (eg convert farms to fish ponds).

Protect – defend vulnerable areas, especially population centres, economic activities, and natural resources.

In addition, authorities will be confronted with a multitude of protection options, including beach nourishment, island rising and the building of dams, dikes and sea walls.

Unfortunately it is not possible to implement all these features into a global model. The resulting model would be too complex. The results presented in this chapter are again based on a much simplified representation of the real world. The model employed is described in detail in Appendix 2 (see also Fankhauser, 1994b). Its basic properties are as follows.

As a major simplification the global analysis presented in this chapter abstracts from the IPCC's 'accommodate' option. That is, coastlines are either protected, or, if this is not done, they will eventually have to be abandoned and will be lost to the rising sea. We suppose that only one protection measure is available per region. Equivalently, we could say that the same measure is cost-effective throughout a region.[8] For example, we can think that each region is protecting its coastlines through a sea wall.

Two kinds of land are distinguished, dryland and wetland. We assume that coasts are protected in accordance to their dryland value, ie more valuable dryland is protected first. As a consequence, the average value of lost dryland depends on the amount of land protected, and decreases with increasing protection efforts. The more land we protect, the lower will be the average value of the remaining, unprotected area. Increased protection will therefore be reflected in the costs of dryland loss both via a quantity and a value effect.

Wetlands, on the other hand, cannot be protected directly. There is nevertheless an indirect link to the amount of coastal protection, in that the ability of wetlands to adjust and migrate inland will be affected by the existence of sea level rise protection measures. As with many ecosystems, it is not so much the absolute level of climate change (or SLR) as the rate of change which poses a threat. Wetlands are in principle able to adjust to SLR by migrating inland. They may, however, be inhibited from doing so by artificial barriers such as a sea wall. The loss in wetlands is therefore inversely related to the degree of coastal protection. The more compre-

---

8 To allow for different measures within a region, eg for different types of coastline, we could further subdivide a region according to coastal types, eg into beaches, open coasts, etc. The same analysis would then have to be carried out for each of them. This will be the method used in the following section on p 81.

hensive the defence measures, the more coastal wetlands will be lost. In addition, wetlands may also be lost if the rate of SLR is too high for a full adjustment.

For lack of data, the model abstracts from the impacts of salt-water intrusion and the costs of increased storm and flood damage. We also assume that the amount of SLR is known with certainty (for a treatment of uncertainty see Yohe, 1991). For each region (or type of coast) considered, the costs of SLR thus consist of three elements: protection costs, dryland loss and wetland loss. We seek to minimize the discounted sum of these three cost streams over time. Costs are minimized with respect to the percentage of coasts protected.

SLR will occur gradually over time. Local authorities are therefore faced with a second problem, *viz* the question of when a sea wall should optimally be built. This problem is of less interest here. Given our model assumptions it will be optimal to simply build defence measures at the same pace as the sea level rises (see Appendix 2). Although simplistic, this result will suffice for our purpose.[9] The concern here is less about the optimal *height* of protection measures than about their *length* and the resulting costs. This is discussed below.

The optimal length of coastal protection

The optimal degree of coastal protection is represented by the variable L^*, which denotes the percentage of coastlines protected. L^* can take any value between 0, if no coasts are protected, and 1, if protection is complete. In Appendix 2 we derive a very simple expression for L^*:

$$L^* = \begin{cases} 0 & if \ L^{opt} < 0 \\ L^{opt} & otherwise \end{cases} \tag{6}$$

with

$$L^{opt} = 1 - \frac{1}{2}\left(\frac{PC^{pv} + WG^{pv}}{DL^{pv}}\right) \tag{7}$$

where the newly introduced symbols have the following meaning. PC^{pv} denotes the present value protection costs under the assumption that all coasts are defended, ie PC^{pv} is the discounted sum of future protection expenditure for the case $L = 1$. Similarly, DL^{pv} denotes the present value of future dryland loss damage if no coasts were protected at all, ie if $L = 0$. WG^{pv}, finally, is the (present value) gain from the inland migration of

---

9 The result is mainly an artefact of the simple protection cost formulation (see Appendix 2). In reality sea walls may be altered less frequently than implied by the model, due to fixed costs (eg hiring and firing costs). The occurrence of storm surges, ie the possibility that the sea level may occasionally rise above its trend level, may justify a further increase in the height of sea walls (see den Elzen and Rotmans, 1992). Uncertainty about future sea levels in general may give rise to precautionary protection.

wetlands under $L = 0$, ie if unrestrained landward migration is possible. Alternatively, WG^{pv} can be interpreted as the forgone gain in wetland growth under full protection.

Equation (7) thus has a straightforward interpretation. The numerator of the term in brackets denotes the present value costs under full protection. They consist of the costs of raising the protection wall (PC^{pv}) plus the opportunity costs of forgone wetland gains (WG^{pv}). There is no dryland loss, as protection is complete. The denominator represents the present value costs in the absence of any protection measures, consisting solely of lost dryland (DL^{pv}).

In our model the optimal level of protection is determined by the relative size of the costs of full protection compared to those under full retreat. *The lower the costs of full protection, the larger will be the share of protected coasts. On the other hand, if the (present value) damage from dryland loss is only modest, the degree of protection will be low.* If the costs of full protection are more than twice as high as the costs under full retreat, no coasts will be protected.

Optimal protection as a function of SLR

The defining factors of L^* vary depending on the projected degree of SLR. The present value costs of dryland loss (DL^{pv}), for example, will be higher, the more the sea level rises over time. Since protection measures are raised at the same pace as the sea rises, the same will be true for the stream of protection costs PC^{pv}. The optimal level of coastal protection will therefore also vary for different degrees of SLR. How will a change in projected SLR affect the optimal level of protection L^*?

To answer this question we assume that the sea level rises linearly over time, and reaches a final height S_τ at the end of the time horizon. The optimal level of protection, L^*, can then be expressed as a function of S_τ, and it is possible to analyse the relation between the two variables. This is compatible with the greenhouse policy debate, where the focus usually is on the expected rise at a certain benchmark date (eg 1 metre by 2100).

Differentiating L^* with respect to S_τ yields, after some rearranging:

$$\frac{\partial L^*}{\partial S_\tau} = (1 - L^*)\left[\frac{DL^{pv\prime}}{DL^{pv}} - \frac{PC^{pv\prime} + WG^{pv\prime}}{PC^{pv} + WG^{pv}}\right] \tag{8}$$

(for an interior solution). $PC^{pv\prime}$, $WG^{pv\prime}$ and $DL^{pv\prime}$ denote the first order derivatives of PC^{pv}, WG^{pv} and DL^{pv}, respectively, with respect to S_τ. It is easily shown that the sign of $PC^{pv\prime}$ and $DL^{pv\prime}$ is positive. Protection and dryland costs will be higher, the higher the projected SLR, as mentioned above. WG^{pv} is independent of SLR, at least in the present model specification. That is, $WG^{pv\prime}$ is equal to zero.

The sign of $\partial L^*/\partial S_\tau$ is therefore ambiguous, a result which is consistent with the intuitive points made in the introductory section of this chapter. On the one hand, more protection is warranted under a high SLR, since more land is at threat and the benefits from protection are higher (the MB curve in Figure 5.1 shifts to the right). On the other hand, the costs of protection are also higher, since more SLR requires higher and more massive installations (MC in Figure 5.1 shifts to the left). The answer thus depends on the relative importance of these two effects.

More specifically, equation (8) tells us that *an increase in projected SLR will lead to a higher (lower) level of protection if the percentage rise in costs under full protection – the second term in equation (8) – is smaller (greater) than the percentage rise in costs without any protection – the first term in equation (8).*

Costs as a function of sea level

We can now also define a *cost function* for the damage from sea level rise. As explained above, SLR damage costs, denoted as $V(S_\tau)$, are equal to the cost-minimizing combination of protection costs and damage *per se* (dry- and wetland loss). Analytically, we obtain $V(S_\tau)$ by substituting the optimal values for the two policy variables – degree of protection and annual increments in height – back into expression (1) of Appendix 2. In the following section we will provide numerical simulations of V and estimate SLR damage cost functions for the countries of the OECD.

SIMULATION RESULTS FOR OECD COUNTRIES

Estimation of parameters

Several data sources are available to calibrate the model on real data, most notably IPCC (1990c), but also Titus et al (1991) and Rijsberman (1991). In some cases, values could be taken directly from these sources, in others they had to be adjusted or extrapolated for the present purpose. See Fankhauser (1994b) for details.

When estimating the costs of SLR we have to be aware of the fact that different types of coasts will warrant different measures of protection. IPCC (1990c) has distinguished between four coastal types: cities, harbours, beaches, and open coasts. We will follow this example and calculate the costs for each of these categories in each country. The total costs of SLR in a country will then be the sum of the four categories. Following IPCC (1990c), we assume that beaches will be conserved through beach nourishment, while all other types will be protected by sea dikes.

The relative length of each type of coast is shown in Table 5.1. It was assumed that coastal wetlands only occur along open coasts. Average land values were set at \$2 m/km^2 for open coasts and beaches, and \$5 m/km^2 for wetlands, as in Chapter 3. The land value for cities and harbours was set at \$200 m/km^2, the highest value reported in Rijsberman (1991). This is sufficient to guarantee an almost complete level of protection of these two coastal types. For lack of country-specific data, regional land values were

Table 5.1 Length of coastlines in OECD countries (km)

	Open coasts	Wetlands	Beaches	Cities	Harbours
OECD	196,729	96,526	8,380	3,850	891.9
AUS	52,310	na	500	254	75.6
BEL	170	2	65	31	25.3
CAN	13,660	na	150	94	32.0
DEN	10,800	5,450	300	78	10.2
FIN	11,825	455	100	36	12.0
FRA	5,190	4,619	800	146	44.7
FRG	2,210	2,210	300	94	38.1
GRE	1,230	650	600	60	16.2
ICE	1,990	1,990	0	10	0.7
IRE	160	160	0	52	4.1
ITA	3,120	1,420	800	264	42.9
JAP	3,920	3,920	100	543	199.1
NET	2,585	2,585	225	51	75.9
NZL	14,975	na	100	151	4.6
NOR	165	165	20	48	14.7
POR	750	750	150	37	5.3
SPA	705	705	500	134	39.7
SWE	12,700	173	200	151	21.9
TUR	750	570	100	82	25.8
UK	6,690	3,473	400	392	64.4
US	50,824	33,700	2,970	1,142	138.7

Source: Fankhauser (1994b).

then compiled by multiplying the OECD average with a series of indicators. These included relative wealth (GNP/capita relative to OECD average) for cities and beaches; GNP/land mass (net of wilderness areas) and percentage of urban population in coastal cities for open coasts – the former an indicator of the scarcity of land, the latter an indication of the relative importance of coastal zones; and international tourist revenues per km of coast for beaches to measure the relative importance of tourist beaches.[10]

It should be emphasized that, as with all estimates on global warming damage, the reliability of many of the available data is rather low. Pairing this with the aggregate character of the model, it should be clear that the obtained estimates can only be indicative of the true result. A considerable range of error has to be accounted for. Nevertheless, the figures provide some insights into the underlying incentives for SLR protection. Estimates

10 The indicator is not ideal because (a) it neglects domestic tourism and (b) it also encompasses tourist activities other than beach recreation. It will thus underestimate beach values in countries where domestic tourism is important, such as the US, and overestimate the value in countries where non-beach tourism dominates.

will be provided for the countries of the OECD, for a time horizon of 110 years, i.e. until the year 2100.

The optimal level of protection

The results with respect to the optimal degree of coastal protection are summarized in Figures 5.2 and 5.3. Not surprisingly the highest degree of protection is achieved in cities and harbours, where the protection rate is nearly 100 per cent in all countries and for all SLR scenarios. The value of the land at threat is sufficiently high to justify full protection. The picture is somewhat more dispersed in the case of open coasts and beaches, where the

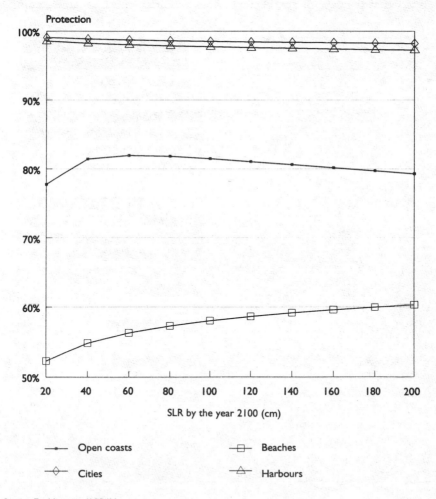

Source: Fankhauser (1994b).

Figure 5.2 Optimal coastal protection in the OECD

optimal level of OECD-wide protection varies between about 75 per cent to 80 per cent and 50 per cent to 60 per cent, respectively (see Figure 5.2).

How will the optimal degree of protection change for different assumptions about SLR? The previous section showed that the sign of the first order derivative of L^* with respect to the expected SLR, $\partial L^*/\partial S_\tau$, is ambiguous and depends on the relative costs of full retreat *vis-à-vis* the costs of full protection (see equation (8)). The numerical simulations now show that in the case of cities and harbours the protection cost element tends to dominate. That is, the optimal level of protection decreases as predictions about SLR increase, although, to be sure, the figures are still not significantly different from 100 per cent even in the worst case. Interestingly this trend is reversed in the case of beaches. Here the potential land loss damage dom-

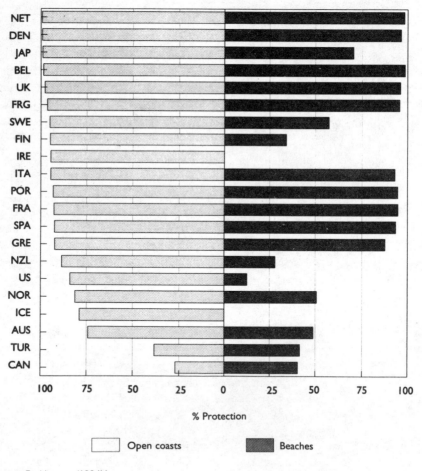

% Protection

Open coasts   Beaches

Source: Fankhauser (1994b).

Figure 5.3 Optimal protection in OECD countries against a 1-metre SLR

inates and the optimal level of protection increases as expectations of SLR become more pessimistic. In a majority of countries the same seems to be true for open coasts. The aggregate picture over the whole OECD, on the other hand, shows a decreasing level of protection at least for higher degrees of SLR (Figure 5.2). This may, however, be an artefact of the poor data quality. No wetland figures were available for the three countries Australia, Canada and New Zealand (see Table 5.1). For these countries the protection cost element therefore dominates, and this also influences the aggregate picture for the entire OECD.

The regional picture can deviate considerably from the OECD average. Figure 5.3 shows countrywise estimates of the optimal protection of beaches and open coasts against a 1-metre SLR. The values for cities and harbours, which are close to 100 per cent for all countries, are not reported in the figure. Regional differences are mainly driven by differences in the value of dry- and wetland. Poorer nations, such as Turkey, and large countries, such as Australia and Canada, tend to have lower protection levels caused by lower land values.[11] The same is true for countries with a low population density, such as Iceland and Norway. In the case of Canada, low land values are further coupled with comparatively high protection costs which leads to optimal protection levels clearly below 50 per cent. Densely populated countries like the Netherlands, on the other hand, have protection values close to 100 per cent even for a low SLR. Low land values are offset by comparatively low costs of protection in Portugal and Ireland, two of the poorest OECD countries. In the case of beaches the protection is increased in countries like Greece and Spain because of the importance of summer tourism in these countries. The low figure for the US, on the other hand, is mainly due to the inaccuracy of the utilized index of beach tourism (see footnote 10).

It should be recalled that these figures are only indicative of the optimal protection levels. As emphasized in the beginning, the design of the optimal SLR protection strategy is genuinely a regional problem which will and should be solved at a local level.

The costs of sea level rise

The results of the numerical simulations of SLR damage cost functions are depicted in Figure 5.4, for the OECD as a whole, and Figure 5.5, for individual countries. The Figures show the costs of SLR as a function of the rise expected by the year 2100. Costs are expressed as the present value of a stream of expenditures over the next 110 years. They differ widely between countries, ranging from less than $10 bn to over $400 bn for a 1 metre rise, although the values are below $50 bn for all but three countries (see Table 5.2). Consistent with the results from Chapter 3 the bulk of damage stems

---

11 Note, that since no wetland figures were available for Australia, Canada and New Zealand, the results overestimate the optimal protection level for these countries.

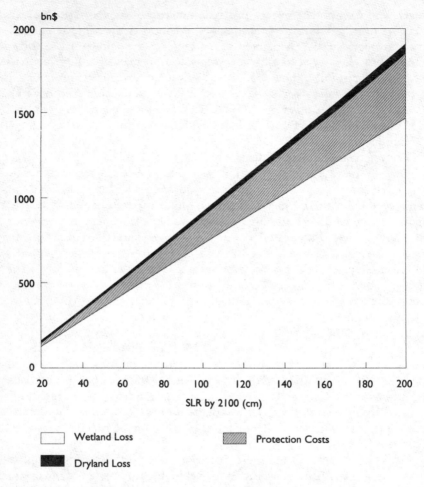

Source: Fankhauser (1994b).

Figure 5.4 The costs of sea level rise in the OECD

from wetland loss. Dryland losses tend to be comparatively moderate and are restricted to low value areas, due to the generally rather high protection levels.

Estimates strongly depend on the choice of the discount rate, as is shown in Figure 5.6. All results presented so far were based on the assumption of zero utility discounting, consistent with Cline (1992a), see Chapter 8. Alternatively if we use the Nordhaus (1992, 1993a, b) assumption of a 3 per cent utility discount rate, costs are reduced by as much as a factor of 3 on average.[12]

12 To be consistent with Cline and Nordhaus the value of the income elasticity of marginal utility also differs between the two cases. It is 1.5 in the low discounting case and 1 under the high rate (see Appendix 1 and Chapter 8).

Table 5.2 Costs of a 1-metre SLR by 2100 in individual countries (present value damage 1990–2100, bn$)

	Total costs	Protection costs	Dryland loss	Wetland loss
OECD	932.47	174.09	27.49	730.89
AUS	34.54	29.55	4.88	na
BEL	0.78	0.61	0.01	0.16
CAN	6.92	3.73	3.12	na
DEN	21.02	7.24	0.05	13.73
FIN	13.05	4.04	0.22	8.79
FRA	32.37	5.70	0.26	26.41
FRG	24.65	2.52	0.07	22.07
GRE	7.85	2.44	0.15	5.27
ICE	1.45	0.84	0.09	0.52
IRE	3.83	0.33	0.00	3.49
ITA	45.27	7.48	0.30	37.49
JAP	141.47	6.83	0.08	134.55
NET	22.41	3.06	0.02	19.32
NZL	16.22	15.03	1.06	na
NOR	9.13	0.43	0.04	8.65
POR	4.70	0.74	0.03	3.93
SPA	31.64	3.24	0.10	28.31
SWE	25.25	8.83	0.42	16.00
TUR	7.51	1.12	0.48	5.90
UK	57.26	7.74	0.14	49.38
US	425.16	62.59	15.96	346.61

Source: Fankhauser (1994b).

The results are also sensitive to the underlying land value parameters, in particular to the value of wetlands. Replacing the $5 m/km^2 central value used so far with the Cline (1992a) estimate of $2.5 m/per km^2 of wetland, for example, leads to a reduction in OECD-wide costs of about $360 bn or almost 40 per cent. Assuming $7.5 m/km^2, the upper bound value given in Titus et al (1991), leads to an increase in costs by a similar percentage.

Damage rises rather rapidly as SLR predictions increase. Figures 5.4 and 5.5 show SLR damage functions which are almost linear. Although it is usually assumed that damage is a convex function of climate change this should not surprise us. Recall that the functions in Figures 5.4 and 5.5 are minimum functions. As such they necessarily have to be less convex than the fixed-policy-response functions which they envelop, and on which convexity predictions are usually based. Further, the main source of non-linearity is typically the construction costs of sea walls, which rise more than linearly with the required height, and thus with SLR. With respect to land loss, the relationship is more likely to be linear, and these costs, particularly wetland damages, dominate the total.

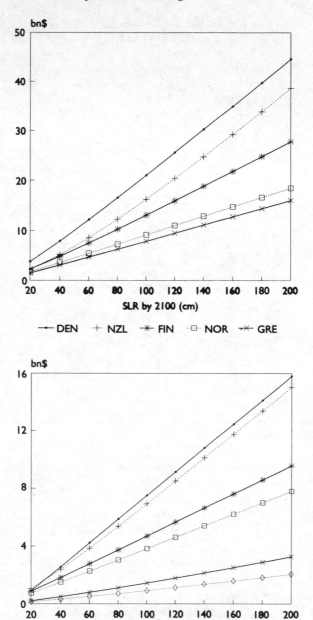

Source: Fankhauser (1994b).

Figure 5.5   The costs of sea level rise in individual OECD countries

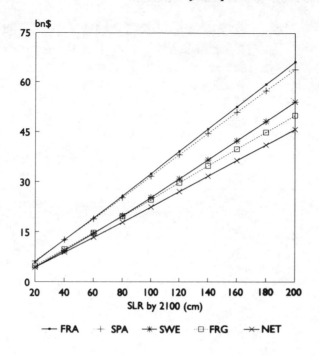

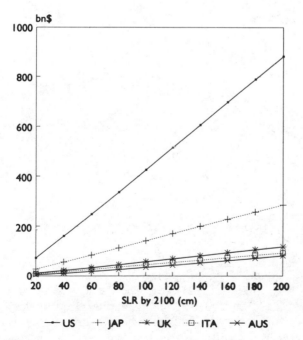

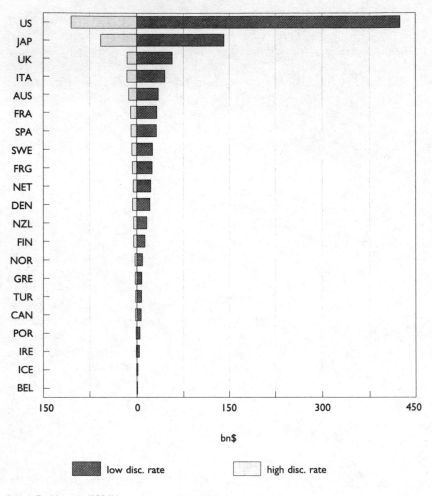

Source: Fankhauser (1994b).

Figure 5.6 The costs of a 1-metre SLR – the impact of discounting

SUMMARY AND CONCLUSIONS

The aim of this chapter was twofold. First we wanted to exemplify the relative role of adaptation expenditures within the costs of greenhouse damage in general, and for SLR in particular. We derived a rule of thumb to estimate the socially optimal degree of coastal protection. It was shown that the optimal level of protection is determined by the ratio of costs under full protection against those under full retreat. The larger the protection costs, the lower will be the degree of protection. The lower the damage resulting if no action were taken, the lower the degree of protection.

In our numerical simulations we found that the optimal degree of protection will vary between about 50 per cent and 80 per cent for open coasts and beaches, depending on the underlying SLR scenario. Cities and harbours are almost invariably protected to the full. A higher SLR will generally lead to a higher degree of protection for beaches, and to a lower protection level for cities, harbours and probably open coasts. Large and sparsely populated countries, as well as poorer nations, will protect their coasts to a lesser degree. They do so not for lack of funds or because of institutional failures, but because more protection is not cost-effective. Resources are more fruitfully invested elsewhere. Albeit only illustrative, the figures suggest that on the whole it may be worthwhile for the wealthy OECD nations to protect the majority of their coasts.

The second aim of Chapter 5 was to estimate the costs of SLR. The aim was to introduce a model which can provide a rough global assessment of SLR damage, to complement the growing number of more accurate, but geographically limited, local case studies. We found that damage costs rise steeply, and almost linearly with the degree of SLR predicted for the end of the planning horizon. By far the most important damage category is the loss of wetlands. The results are, however, quite sensitive to several of the underlying parameters. It should also be clear that SLR costs to the countries of the OECD are only a fraction of total world-wide greenhouse damages (see Table 3.15 above). No conclusions can therefore be drawn on the optimal level of greenhouse gas abatement, based on this damage alone.

A high level of aggregation allowed, and indeed was necessary for, a global assessment of SLR damage. The analysis would be enhanced by incorporating more regional detail about the geographical and socio-economic differences in individual countries. Recall from Chapter 3 that the value of land is only an imperfect indicator of the true welfare loss to consumers, and for many people their homeland may be worth more than just its market value. Further, the resettlement of people from abandoned areas may not take place without friction and may be subject to considerable adjustment costs (see Pearce, 1993b). The costs of resettlement of a climate refugee (excluding the disutility from hardship) have been assessed at about $4,500 per person (see Chapter 3). These were neglected here, and the figures presented are therefore likely to underestimate the true costs.

The costs of sea level rise are a major part of expected climate change damage, even if a cost-efficient policy response is taken. But coastal problems would also be significant in the absence of climate-induced sea level rise. Increased human settlements and economic activities in coastal zones have created an augmented pressure on natural coastal ecosystems, affecting both the diversity and stability of coastal habitats. Human interference in the physical appearance of shorelines, the construction of levees and dams, and upstream activities like the discharge of effluent into the water have led to a world-wide destruction of coral reefs, mangroves, sea grasses and salt marshes (Turner et al, 1994b). Climate-induced sea level rise merely adds an additional dimension to an already pressing environmental problem.

PART III
Global Warming Policy

———— ◆ ————

Chapter 6

The Costs of Greenhouse Gas Abatement

ENERGY MODELS AND CARBON ABATEMENT

Economists have been interested in the question of greenhouse gas emissions for quite some time. However, the interest of the early models has not so much been in the costs of greenhouse gas abatement, as in the construction of future business-as-usual emission trajectories. That is, the focus was still on the question of whether a greenhouse problem existed in the first place, rather than on the evaluation of policies towards its prevention. This changed in the aftermath of the Toronto conference of 1988, when the emphasis shifted from emission prediction to abatement cost assessment. In the early 1990s the number of models began to mushroom. In many cases existing models were extended to include a carbon module, in others altogether new models were built from scratch. Today there exist greenhouse gas abatement models for almost every country of the OECD, in addition to a few developing country studies and several global models.

Types of models

Modelling attempts can be, and have been, classified in several ways and according to different criteria (see eg UNEP, 1992). An important distinction, both in the light of the different underlying philosophy and the difference in results, is between the so called *bottom up* and *top down* approaches. Top down models can then be further divided into macro-economic and resource allocation or equilibrium models. For all categories there are global models as well as country studies.

Bottom up models are technology oriented, engineering based studies, which concentrate on the availability and performance of individual energy supply technologies. The menu of available technologies is summarized in technology cost curves, a list of available technologies ordered according to the costs of emissions reduction, which are then compared to the demand for energy services.[1] The costs of reducing CO_2 emissions are typically small or even negative, since the most efficient technology is often found to provide the required services at a lower cost than the measures currently in

---

[1] For a detailed description of the approach see eg Johansson and Swisher (1994).

place. Examples of bottom up studies include Mills et al (1991), COHERENCE (1991), and Chandler (1990). Examples for developing country studies can be found in Pachauri and Bhandari (1992).

Top down models, on the other hand, are based on an economic approach in which energy is treated as one of several production inputs in a model of production and economic behaviour. By assumption production is efficient, so there is no scope for 'free lunches'. Macroeconomic models, the first main type of top down models, concentrate on the short run and on the disequilibrium/adjustment effects occurring after the imposition of policy measures, and may thus best explain short-term, transitional effects of carbon abatement on macroeconomic indicators like inflation, employment, interest rates, etc. Models are usually empirically tested and thus based on historically observed trends. Macroeconomic models utilized in the carbon debate include the HERMES model used by the European Commission (see eg Standaert, 1992; Karadeloglou, 1992), two US models – DRI (Brinner et al, 1991) and LINK (Kaufmann et al, 1991) – and the Cambridge MDM for the UK (Multisectoral Dynamic Model – see Barker et al, 1993).[2] A global macro model distinguishing eight regions is Hall et al (1994).

The other main category of top down models are resource allocation models. These are long-run equilibrium models in which changes in relative prices are the main factor to clear markets in the longer term. Models can either be of the comprehensive general equilibrium type or are partial equilibrium models restricted to one sector, usually energy. Examples of the latter type are the IEA model (Vouyoukas, 1993) and the well known and often used Edmonds and Reilly model – a detailed, long-term global energy model distinguishing nine geopolitical regions, three greenhouse gases and six primary sources of energy (see Edmonds and Reilly, 1983; Edmonds and Barns, 1991). Another well-known global model is Global 2100 (Manne and Richels, 1992), which combines a partial equilibrium energy module with a macroeconomic model (see Box 6.1).

Rather than concentrating on one specific sector, general equilibrium models (or CGE – computable general equilibrium – models) emphasize the interconnections between the different sectors of an economy and consider several markets simultaneously. CGE models are driven by changes in the relative prices of goods, caused for example by the introduction of a carbon tax. A long-run equilibrium is achieved at a set of prices which clears every market in the economy, that is supply equals demand for all commodities. Examples of general equilibrium models abound, and include prominent carbon abatement models, such as the global models GREEN (OECD, 1992c), Whalley and Wigle (1993; 1991a, b), McKibbin and Wilcoxen (1992a, b), and the Second Generation Model by Edmonds et al (1992),

---

2 Most models are described in several publications. Reference is usually made to the most recent or most comprehensive one.

Box 6.1
Three main global abatement cost models

GREEN

Probably the most comprehensive carbon abatement model available at the moment is the GREEN (GeneRal Equilibrium ENvironmental) model developed at the OECD. GREEN is a dynamic, general equilibrium model of the global economy. It includes 12 regional submodels and distinguishes between 11 economic sectors. All regions are linked through trade flows. The model is recursive, ie saving decisions affect future output through the accumulation of capital, but the investment decisions of firms are not directly modelled. Capital accumulation is modelled in a putty/semi-putty fashion, ie there are some factor market rigidities and capital is partially sector specific (for details see Burniaux et al, 1992b).

Global 2100

Another main global model is Global 2100, described in Manne and Richels (1992). Global 2100 is a dynamic non-linear optimization model which divides the world into five geopolitical regions: the US, all other OECD countries, the former Soviet Union, China, and the rest of the world. The model has less sectoral detail than GREEN, and trade flows are not fully accounted for. On the positive side the model is based on forward-looking behaviour, and the time horizon is extended to 2100 (2050 in GREEN). Global 2100 has been the basis for several other modelling attempts, including CRTM (the Carbon Rights Trade Model) by Rutherford (1993), which incorporates trade links although at the expense of the perfect foresight assumption, and CETA (Peck and Teisberg, 1992, 1993a, b), which extends Global 2100 to include a greenhouse damage sector.

The PNL Second Generation Model

A third major global model is currently produced at the Pacific Northwest Laboratory. The Second Generation Model (SGM) is a detailed computable general equilibrium model based on the earlier Edmonds and Reilly (1983) energy model. It will cover all main sources of greenhouse gases, including non-energy sources, and contain sectoral detail on households, government, agriculture, energy and other products. Trade will occur in all of the three latter sectors. In its initial version SGM will distinguish 11 geopolitical regions. It is a behavioural model (as opposed to an optimizing model, such as Global 2100), which will allow to test for different expectation formation hypotheses (see Edmonds et al, 1992).

---

which is currently under construction (see Box 6.1). National general equilibrium models include Jorgenson and Wilcoxen (1990, 1992) and Goulder (1993) for the US, Blitzer et al (1992) for Egypt, and Glomsrød et al (1992) for Norway, to name only four.

An abatement cost benchmark

As diverse as the number of available approaches, is the variety of model results. Despite this diversity, a general picture has started to emerge from model comparisons and surveys, and it appears that initial abatement efforts could be obtained at very low costs. Further steps would, however, become increasingly costly, at least according to top down models. The burden imposed on the economy will crucially depend on the pace at which measures are phased in. Under a gradual abatement scheme costs are considerably lower since the need for a premature replacement of capital is reduced (see eg Richels and Edmonds, 1993).

An often-used benchmark is that *a 50 per cent cut from baseline emissions by about 2025–50 would reduce GNP by some 1 per cent to 3 per cent* (see Cline, 1994, who arrived at the same figure through a back of the envelope calculation). Tables 6.1 and 6.2 show the costs of emission stabilization and the required carbon tax, as estimated by three top down models. Bottom up models are consistently more optimistic and predict relatively low abatement costs even for substantial cuts. According to this class of models as much as a 20 per cent to 50 per cent reduction in energy demand could be achieved at next to zero extra cost (see Grubb et al, 1994).

Although sizeable in absolute terms, several authors have noted that abatement costs are not necessarily extravagant if seen in an intertemporal context. Schelling (1992) for example notes that a 1 per cent to 3 per cent cut in GNP would merely postpone 'the GNP of 2050 until 2051' (p 8). The rate of economic growth over the next decades would be reduced by less than 0.1 percentage points, 'well below the margin of forecasting error', as Ekins (1994, p 40) observed. Abatement costs will also appear less sizeable once several spill-over effects are taken into account, which will occur as a consequence of carbon abatement. Particularly important are the benefits of reduced air pollution and of a less distortionary tax structure. As a matter of fact, the combination of these two effects could often be sufficient to render carbon abatement worthwhile in its own right (see the section on 'secondary benefits', p 102 and Chapter 7).

Table 6.1   GDP loss due to emission stabilization at 1990 levels (% change relative to baseline)

	Edmonds Reilly		GREEN		Manne Richels	
	2020	2050	2020	2050	2020	2050
USA	0.58	0.81	0.29	0.36	1.08	2.11
other OECD	0.74	0.92	0.30	0.62	0.75	1.31
ex-USSR	0.02	0.33	1.39	2.07	1.34	0.79
China	3.42	5.67	3.37	5.56	2.80	4.05
ROW[a]	1.76	2.96	3.89	4.45	5.20	5.38

a rest of the world

Source: OECD (1993).

Table 6.2 Carbon tax necessary for emission stabilization at 1990 levels ($/tC)

	Edmonds Reilly		GREEN		Manne Richels	
	2020	2050	2020	2050	2020	2050
USA	88	119	65	51	136	208
other OECD	114	134	60	85	121	208
ex-USSR	0	26	62	86	119	30
China	237	478	279	466	270	580
ROW[a]	361	779	255	404	329	709

a rest of the world

Source: OECD (1993).

SHORTCOMINGS AND EXTENSIONS

Model comparisons

Overview papers trying to survey and categorize the host of different models and to compare their results are themselves many.[3] Unfortunately, the comparison between different models is not always easy. Differences in scope and underlying assumptions make it almost impossible to compare like with like. Two model comparison projects which have recently been undertaken at the OECD and at the Stanford Energy Modelling Forum (the EMF-12 project) were able to shed some light onto the working of the different models by standardizing key underlying assumptions, eg with respect to population growth, GNP growth and the imposed carbon constraint. The results are summarized in OECD (1993) and Gaskins and Weyant (1993b), respectively.[4]

Important cost parameters singled out in both comparisons include the elasticity of substitution, both between different fuels and between energy and other factors of production, the price and time of availability of a carbon free backstop, and the flexibility of the capital stock. The more malleable the capital stock, ie the cheaper it is to reallocate capital between industries, the lower will be the costs of swift abatement. Evidently, the costs of abatement will also crucially depend on the type of abatement policy set in place. Most models correspond in this respect and assume that abatement will be prompted through the imposition of a carbon tax.

Energy efficiency improvements and future emission trends

A surprising result which came out of both model comparison projects is

---

3 See eg Grubb et al (1994), UNEP (1992), Hoeller et al (1992, 1991), Boero et al (1991), Cline (1992a), and Nordhaus (1991a).
4 See also Gaskins and Weyant (1993a) and Weyant (1993).

that even with standardized assumptions on GNP growth, models still produce considerably different baseline (BAU) emission paths. The crucial explanatory variable for this appears to be the rate of autonomous energy efficiency improvements (AEEI), a parameter which encompasses all non-price induced improvements in energy efficiency, eg due to technical progress or structural change. A low AEEI value appears to be the main reason for the rather high BAU emission path in, for example, Global 2100. Indeed Manne and Richels have repeatedly been criticized for being too pessimistic about AEEI, particularly by the bottom up school.[5]

The BAU path, and thus AEEI, will not only determine future warming rates, it will also affect the costs of abatement. The higher the unconstrained (BAU) emissions, the more abatement will be necessary to reduce emissions to a given absolute target level.

Unfortunately, there is little empirical evidence about AEEI. Proops et al (1993) use a decomposition procedure and input-output tables for Germany and the UK to show that the historic energy efficiency trend (together with that for carbon intensity) is such that, should it continue, carbon emissions of these two countries may stabilize almost naturally over time. A similar decomposition with a wider set of countries was also carried out by Ogawa (1991). Both studies point towards a high value of AEEI. Low and possibly even negative values are, on the other hand, reported in an econometric study by Hogan and Jorgenson (1991). The question, of course, is to what extent historic trends, often observed over a few years only, can be extrapolated over decades into the future to give an indication about AEEI in the very long run. Another question which remains is how far a strong greenhouse policy will itself trigger technological change, eg through increased R&D. Progress would then not be autonomous, but endogenous. A recent poll of experts conducted by Manne and Richels (1994a) has produced expected future AEEI figures of about 0.7 per cent per annum, slightly higher than their initial estimation.

Existing energy taxes and subsidies

The costs of greenhouse gas abatement also depend on already existing distortions in the energy market, eg due to trade barriers or the existence of cartels and monopolies. Most notably, distortions occur in the form of either energy taxes or subsidies. From economic theory it is well known that a tax put on top of an already existing one will cause higher costs than a tax imposed on an undistorted market. In the case of a linear demand curve the deadweight loss from taxation rises quadratically with the tax rate (Clarke, 1993; Burniaux et al, 1992a). The presence of already implemented energy taxes could thus considerably affect the costs associated with global warming

---

5 Eg Williams (1990), Wilson and Swisher (1993), and Grubb et al (1994).

abatement. This, of course, is under the proviso that existing taxes are not there to correct for other externalities like air pollution.

The size of existing energy taxes and the question of how they could be rearranged to account better for the carbon externality has been studied by Hoeller and Coppel (1992). They estimate that existing energy taxes in OECD countries correspond to an implicit average carbon tax of about \$70/ tC. Replacing the current tax structure with a pure carbon tax of this size would reduce OECD-wide carbon emissions by about 12 per cent. Since the measure would result in a less distortionary tax structure, this abatement would actually be achieved at a negative cost.[6] Interestingly, the scheme would mostly boost oil, the most heavily taxed fuel under the current regime, and not the cleaner gas. The average price of oil would fall by 17 per cent, whereas gas and coal prices would increase by 17 per cent and 77 per cent, respectively.

In a world-wide context it has been argued by Larsen and Shah (1992), among others, that the concern should be less about already existing taxes than about energy subsidies. They estimate that energy subsidies are in the order of \$230 bn world-wide, a figure which is roughly confirmed by Burniaux et al (1992a). Their removal would then give rise to a massive free lunch in carbon abatement, an option policy-makers should clearly make use of.

Revenue recycling

A related question is that of revenue recycling. If abatement is achieved through a carbon tax, governments will be able to acquire substantial revenues, and the question is how these should be utilized. It has been argued by, for example, Pearce (1991) that if revenues were used to replace existing taxes, such as those on income, the resulting overall reduction in tax distortions would constitute a 'double dividend' of potentially considerable size. Taxing an activity like income has the disadvantage that it creates a disincentive to work and biases people's preferences towards the untaxed activity leisure. Replacing an income tax with a carbon tax can help to correct this distortion. Nordhaus (1993a) has estimated that the inclusion of double dividends would justify a carbon tax of about an order of magnitude larger than would otherwise be the case. The question thus clearly deserves further attention. We will come back to it in Chapter 7.

Top down vs bottom up models: closing the gap

The main future challenge to the abatement cost literature is to close the gap between the results from top down and bottom up studies. First steps in this direction are found in Grubb et al (1994), Grubb (1990) and, from a distinct

---

6 Unfortunately, the study ignores the impact of the change of regime on other environmental problems such as acid rain. See 'Secondary Benefits', p 102, below.

bottom up point of view, Wilson and Swisher (1993). A key reason for the difference in results, put forward by Grubb et al, is the possible existence of hidden costs not accounted for in bottom up models. These may, for example, include costs of installation or the premature replacement of old devices, or else a new technology may simply not be as perfect a substitute as initially thought. Bottom up models may thus tend to overestimate the scope for cheap abatement options.

On the other hand, the energy market may not work as efficiently as top down models assume. Inefficiencies may occur as a consequence of market or information failures, for example, or because of budget constraints faced by poor households: restricted access to credit markets may prevent poor households from investing in energy efficiency improvements (Grubb, 1990; Brechling and Smith, 1994). Brechling and Smith (1994) have found some evidence of such imperfections, in particular of what they call the tenure effect. Because landlords are unable to fully capture the benefits from energy efficiency investments through higher rents, private rented properties tend to have lower energy efficiency standards than owner-occupied homes. Clearly, the removal of such inefficiencies, in as far as they exist, would give rise to cost-free abatement opportunities which are not incorporated in top down assessments.

These initial considerations, then, seem to imply that the true costs of abatement lie somewhere between the top down and bottom up estimates, and that differences are mainly the consequence of overlooked aspects on both sides. Arguably, the discrepancies may also be deeper, however, and may essentially have to do with a fundamentally different understanding of the whole issue in the two schools (Grubb et al, 1994). A first analytical attempt to combine the top down and bottom up philosophies in a single model is MARKAL-MACRO (see Manne and Wene, 1992, as quoted in Wene, 1993). Further attempts in the same direction are clearly required.

SECONDARY BENEFITS

The (net) costs of greenhouse gas abatement may also be affected by the existence of ancillary, or so-called *secondary*, benefits. As was already mentioned in Chapter 3, the benefits of greenhouse gas abatement may not be limited to reduced climate change costs alone. Substantial benefits could also occur in the form of local or regional air quality improvements and reduced traffic externalities like accidents or congestion. Transboundary problems like acid rain apart, ancillary benefits will predominantly arise in the abating countries themselves, and will thus reduce their net costs of abatement.

The issue of air quality is tied to global warming in that both problems are caused by largely the same activities, in particular the combustion of fossil fuels. Because no economical CO_2-removal technologies currently exist, attempts to limit CO_2 emissions will by and large concentrate on cutting down the use of fossil fuels. A reduction in CO_2 emissions will therefore also reduce the emission of classic air pollutants like SO_2, NO_x and particulates. The secondary benefits from improved air quality may be quite large (Pearce, 1992), and their measurement has gained increasing attention recently. In addition to the back of the envelope calculations by Pearce (1992) and Ayres and Walter (1991) the issue has been addressed in particular in the three related Norwegian studies by Glomsrød et al (1992) and Alfsen et al (1993, 1992). Estimates further exist for the UK (Barker, 1993) and the US (Scheraga and Leary, 1994). Amano (1994) has extended the Alfsen et al (1992) estimates to several Asian regions, using the same benefit-abatement ratios (air quality benefits, in per cent of GDP, per percentage cut in air pollution). Table 6.3 provides an overview of several estimates from these studies. It shows percentage reductions in air pollution levels ranging from 1 per cent to over 30 per cent, depending on pollutant and region. Note that while the underlying abatement scenarios differ from study to study, they all roughly succeed in stabilizing CO_2 emissions at about 1990 levels.

What is the value of these emission reductions? Estimates which measure the social costs of each pollutant vary widely between regions, depending on local factors like baseline air quality, ecosystem vulnerability, population at risk etc. Figures may also differ because impacts caused by a combination of gases are attributed to the initial sources in different ways. A selection of illustrative results is given in Table 6.4. On the basis of such estimates the secondary benefits of CO_2 emission stabilization amount to several billion dollars in each of the countries analysed. The figures are rather high, compared to the initial abatement costs. This is particularly the case for the United Kingdom for which Pearce (1992) and Barker (1993) both observe that 'it seems very likely that the [secondary] benefits are far greater than the conventional GDP costs' (Barker, 1993, p 19). In Norway, secondary benefits amount to $40–$140 per tonne of carbon abated, and could offset about one third of the initial GDP costs (Alfsen et al, 1992). A similar picture emerges in Asia (Amano, 1994).[7] In the case of Japan and India, estimated secondary benefits exceed the primary costs of stabilizing emissions at 1990 levels. For China and the group of 'Dynamic Asian Economies', secondary benefits are estimated to offset about one-third of the initial abatement costs. In many cases, greenhouse gas abatement could thus turn out to be a virtual 'no regret' option, that is a measure worthwhile pursuing in its own right, independent of global warming considerations.

An alternative way to measure secondary benefits is by estimating the change in the costs of meeting air quality standards. Many industrialized countries are committed to significant cuts in the emission of air pollutants.

---

7 A big caveat is however necessary with respect to these results, in that the Norwegian benefit/ abatement ratios used in the study may only poorly represent those observed in Asian countries.

Table 6.3 Reduced air pollution due to CO_2 abatement (% reduction compared to baseline)

Country	Year	Policy/Scenario	CO_2	SO_2	NO_x	CO	Partic.	VOC	Secondary Benefits ($/tC)	Sources
Europe[1]	2000	EU carbon/energy tax								
		– current product. structure	9.4	7.4	6.2	na	na	na	6.1	Alfsen et al (1993)
		– cost efficient regime	9.7	9.3	6.4	na	na	na	6.6	
Norway	2000	Emission stabilization (at 1989 level)	15.0	20.8	10.8	24.1	4.3	na	40–140[2]	Alfsen et al (1992)
UK	2005	EU carbon/energy tax	12.1	38.3	10.6	9.6	30.3	1.1	40–1040	Barker (1993)
US	2000	Emission stabilization								
		– through carbon tax	8.6	1.9[3]	6.6	1.5	1.0/1.8[4]	1.4	2.0–20	Scheraga and Leary (1994)
		– through Btu tax	8.6	2.2[3]	6.6	3.4	1.6/2.2[4]	2.7	3.5–28	

1 Western and Eastern Europe (UN ECE region). Tax in six EU countries (France, Germany, UK, Italy, Netherlands, Denmark) and Scandinavia (Norway, Finland, Sweden) only

2 including road traffic benefits (reduced congestion, noise, accident, and road damage costs)

3 SO_x

4 PM_{10}/TSP

Sources: as shown.

Table 6.4 The social costs of air pollution ($/tonne)

Country	SO_2	NO_x	CO	Particulates	VOC	Source
UK	367	124	15	21,333	na	Pearce (1994b)
UN ECE[a]	637	490	na	21,333	na	Pearce (1994b)
Norway	500–7,600	1,600–31,400	1–13	2,100–27,700	na	Alfsen et al (1992)
US	4,800	2,000	na	2,700	na	PACE (1990)
US	300–1,800	10–100	na	400–10,900	360–2,400	Scheraga and Leary (1994)

a Damage done by a tonne of *UK emissions* to the whole of Western and Eastern Europe (incl. UK)

Sources: as indicated.

The Helsinki protocol of 1985 requires a cut in sulphur emissions of 30 per cent by 1993, compared to 1980 levels, for selected European countries. Under the 'Second Sulphur Protocol' further reductions will be required up to 2010. The Sofia Protocol on nitrogen dioxide commits signatories to a freeze on emissions at 1987 levels. Greenhouse gas abatement will lower the amount of traditional air pollution abatement needed to meet these targets. Alfsen et al (1993) calculate that the introduction of a carbon/energy tax would reduce the expenses for traditional SO_2 and NO_x abatement in the nine countries introducing the tax by 25 per cent to 30 per cent and 12 per cent to 25 per cent, respectively.

The question of secondary benefits from carbon abatement is qualitatively different from the more relevant one about the optimal abatement mix with respect to all externalities. The secondary benefit argument suffers somewhat from an implicit primacy of the greenhouse problem, in that improvements in other areas are greeted as welcome side-effects of a global warming policy, but are not considered or sought in their own right. This is not the ideal way to proceed, especially given the relative size of air pollution benefits. In an ideal world, each pollutant would be taxed in proportion to the environmental damage it causes. If there are interdependencies between them, as is the case with global warming and air pollution, this may affect the relative tax levels, and the problems should thus be considered *simultaneously* (see Ingham et al, 1993).[8] Initial work in this direction has been carried out by Bergman (1991), but without paying sufficient attention to the exact nature of the interdependencies involved. Of course, taxing each pollutant separately may not always be feasible. A related question is therefore that of multiple externalities in the presence of only one policy instrument. A

---

8 Connections between global warming and air pollution exist on both the damage and the abatement side, see Chapter 3. With respect to the former, atmospheric sulphur aerosols will dampen (or 'mask') global warming, while a warmer temperature will aggravate air pollution. On the abatement side it is usually assumed that CO_2 abatement will have a positive impact on the emission levels of air pollutants (the secondary benefit argument), while sulphur abatement, if achieved through the installation of end of pipe scrubbers, may lead to an increase in CO_2 emissions.

carbon tax, which is the most efficient way to deal with the global warming externality (see Chapter 7), may then no longer be the optimal policy instrument.

Chapter 7

Policy Instruments and Carbon Tax

INTRODUCTION

Compared to the issue of carbon abatement costs, the question of policy instruments is relatively uncontroversial. Although more policy-oriented authors argue for a 'mixed bag' containing a variety of measures, the economic literature clearly favours the use of market-based instruments like pollution taxes or tradeable emission permits. It does this for the well-known reasons of static and dynamic efficiency (eg Smith, 1992b). Taxes and permits allow the achievement of a given emission target at minimum economic costs (static efficiency), and in addition they create a dynamic incentive for the development of cheaper abatement techniques in the long run (dynamic efficiency). The debate thus soon concentrated on the relative merits of taxes versus permits, and on the question of the most appropriate tax base. Taxes could be levied on the market value of fuels (an *ad valorem* tax), on their energy content (an energy tax) or on their carbon content (a carbon tax). They could be imposed on the consumer side or on the producer side. Further questions concern the problem of already existing distortions in the energy market, the distributional consequences of a tax, and the use of the collected tax revenues.

CHOOSING THE INSTRUMENT

Taxes vs permits

According to a well-known paper by Weitzman (1974), a tax system is more desirable than a tradeable permits system if the slope of the marginal benefit curve is greater than that of the marginal abatement cost curve and there is *uncertainty* with respect to the costs of abatement. Peck and Teisberg (1993a) have argued that for global warming this may be the case, implying that taxes should be preferred over permits. In an international context Berger et al (1992) could show that under *imperfect competition* in the market for fossil fuels, a permits system will result in higher producer prices than an international carbon tax. A tax may also be preferable if individual countries are large enough to influence the permits market (see Hoel, 1991a). In most analyses, however, taxes and permits are assumed to be equivalent and the choice is then determined by analytical convenience or by the preference of the author.[1] Tradeable permits have so far mainly been analysed in the context of international cooperation between countries (eg UNCTAD, 1992; see also Chapter 10). In the domestic debate taxes have gained more prominence.

Fine tuning the tax option

The imposition of greenhouse taxes of some sort is now discussed in several countries, most importantly the United States and the countries of the European Union. Carbon or energy taxes are already implemented in most Scandinavian countries as well as in the Netherlands. If the choice is of a tax, economic efficiency would require that it be levied as closely to the polluting source as possible. In the case of global warming this would mean a carbon tax, since CO_2 emissions are more or less proportional to the carbon content of a fuel. An energy, and particularly an *ad valorem*, tax would be less closely correlated with the pollution potential of a fuel (Leary and Scheraga, 1993). The difference, basically, is that a carbon tax would lead to a larger change in the relative price between fuels and would thus give stronger incentives for interfuel substitution between dirty and clean alternatives. Analyses of the cost difference between the different tax options include Jorgenson and Wilcoxen (1992) and, in the context of the EU carbon *cum* energy tax, Manne and Richels (1993) and Karadeloglou (1992). They confirm that the choice of the tax base can make a noticeable difference in terms of costs and the tax level required. Jorgenson and Wilcoxen (1992), for example, calculate that for the US the cost of stabilizing carbon emissions by means of an *ad valorem* tax would be about twice that of achieving the same goal through a carbon tax. An energy (Btu) tax would cost about 20 per cent more than the carbon option.

The difference between a producer- and a consumer-based tax mainly plays a role in an international context due to trade in fossil fuels.[2] Under a production-based tax, revenues flow to fuel exporting countries, while under a consumption-based tax they remain in the importing country. Whalley and Wigle (1991a, b) estimate that, not surprisingly, the difference between the two concepts is particularly important for oil exporting countries, but also for large importers like the EC and Japan. The problem is mainly one of distribution, though. The welfare costs for the world as a whole are almost identical for the two schemes, at least under full participation.

Another, more theoretical, discussion has focused on the optimal inter-

---

1 For a discussion of the tax option see eg OECD (1992a), or Pearce (1991). Tradeable permits are discussed in UNCTAD (1992) and OECD (1992b).

2 Although see Scheraga and Leary (1992) for domestic considerations. If a certain type of fuel can be converted into another form with a different carbon content (eg coal into synthetic gas), the impact of a carbon tax will differ depending on whether it is levied before or after the conversion.

temporal path of a carbon tax. In dynamic optimization models an increasing marginal damage over time typically also leads to a rising tax rate over time (see Chapter 4). This view has been challenged by Kverndokk (1994), Sinclair (1992) and Ulph et al (1991), with an argument based on the theory of exhaustible resources. They argue that as fossil fuels are gradually exhausted, the tax rate will eventually have to start decreasing. The tax path should therefore be hump-shaped, with an initially rising tax rate which is falling again in later years. However, given the abundance particularly of coal reserves, it seems unlikely that the exhaustion constraint will start biting in the foreseeable future.

Joint implementation

While tax solutions have emerged as the main alternative to command and control measures on a national level, there appear to be signs of a gradual endorsement of the permits idea in the international context. It seems somewhat unrealistic to envisage the emergence of an international carbon tax scheme, in which countries would pay conceivably large amounts of money to an international greenhouse agency in proportion to their greenhouse gas emissions.[3] A permit solution could therefore be more attractive. On the other hand, the crucial question of emission property rights, ie the initial distribution of pollution entitlements between parties, is similarly fraught with political difficulties. An alternative to a pure permit system has therefore recently gained prominence: the idea of joint implementation, or carbon offsets – an earlier term for basically the same concept, although mainly used in the context of carbon sequestration.[4] The option is explicitly included in the Climate Convention, and demonstration projects are already under way, eg between Norway, on the one hand, and Mexico and Poland, on the other.

Joint implementation schemes essentially enable high abatement cost countries to be credited for (cheaper) abatement or sequestration efforts undertaken abroad. The idea is thus similar in spirit to that of tradeable emission permits. Country A undertaking abatement in country B under a joint implementation scheme is conceptually similar to country B selling emission permits to country A. Under certain circumstances the two systems are indeed identical, as was pointed out by Bohm (1994).

Two conditions have to be met for this to be the case. First, the scheme set in place has to be multilateral and involve a sufficiently large number of countries. This is primarily necessary to guarantee the emergence of a reasonably well-functioning market in abatement projects. Compared to a scheme based on bilateral negotiations it would also help to reduce the otherwise massive transaction costs. For some offset trades in the US,

---

3 See Hoel (1992). A contrary view is Martin et al (1992).
4 For surveys see eg Jones (1994), Bohm (1994), and earlier Dudek and LeBlanc (1990).

administrative costs amounted to almost 1 per cent of the total budget, and were 'perceived as being a major impediment to trading' (Barrett, 1993a, p 12). In the pilot projects between Norway, Mexico and Poland, administrative costs were even higher, not least due perhaps to their character as demonstration projects.

The second condition is that there has to be a clearly defined joint emission target. Practically, this means that all countries involved have to be committed to a certain emissions policy. In the absence of such commitments the calculation of the incremental abatement achieved through a project becomes extremely difficult. The main problem is that emissions in the baseline scenario without a project cannot be observed. It is very difficult to prove that a forest area which is protected through joint implementation would indeed have been deforested otherwise, and that no other area was deforested in its place. Host countries have a clear incentive to declare projects as joint implementation activities which would have been undertaken anyway. Similarly, donor countries are interested in overstating the incremental abatement achieved in a given project. Ironically, these problems are the more imminent the more attractive a project, since the projects which offer the cheapest abatement options are also those which are most likely to be economically viable in their own right.

The discussion would thus suggest that the most successful projects would be multilateral schemes between developed countries (which are more or less all committed to emission stabilization, see Chapter 1). Yet, in the political discussion so far, the emphasis has been more on bilateral projects between developed and developing countries, perhaps with the Global Environment Facility acting as a clearing house. The value of this type of deal is less well understood and a lot of mainly practical questions have yet to be answered (see eg Jones, 1994).

THE DISTRIBUTIONAL CONSEQUENCES OF A CARBON TAX

Probably of more political relevance than the choice of the appropriate policy instrument is the question about the distributional consequences of policy measures. Distributional issues arise in several respects, including the impacts on different countries, income groups and economic sectors. The first aspect has probably been the most controversial so far, and has dominated the discussion about international cooperation towards a greenhouse protocol. We will come back to it in Chapter 10.

Income distribution

Income aspects have been analysed by Alfsen et al (1992), Poterba (1991), Pearson (1992), Symons et al (1994) and in several papers by Smith (eg 1994; 1992a). The distributional incidence in developing countries is considered in Shah and Larsen (1992). Most of these studies confirm the initial intuition that a carbon tax may be *regressive,* ie that low income households will be harder hit than high income households. Distributional analyses often take the current spending pattern of households as a starting point, and observe that poor households spend a larger share of their income on domestic energy. Although the picture is softened somewhat through the inclusion of motor fuel expenditures, whose importance increases with income, this is then taken as an indication for the regressive character of a carbon tax. Table 7.1 summarizes the results of such an analysis for seven European countries.

Several effects are neglected in using current spending patterns, however. First, there will be a demand response of households to higher energy prices, ie the total amount of energy consumed will drop. Second, there will be an indirect effect of the tax on the remainder of the consumption bundle. Goods which are energy intensive to produce will, in particular, become more expensive, and this will again affect demand.[5] Both these aspects are assumed to have a dampening effect on the regressiveness of the tax, basically because low income households are expected to react more strongly to price changes, and because high income households tend to consume more goods which are energy intensive to produce.

Distributional assessments may also suffer from measurement errors arising from the use of an incorrect proxy for economic well-being. Poterba (1991) has noted that the most appropriate measure of wealth would be people's lifetime income, and argues that consumption expenditures would be a more reliable indicator of this than the more volatile current income. An expenditure-based analysis again seems to provide a less dramatic picture than one based on current income. On the whole, the regressive impact of a carbon tax is therefore perhaps less severe than often thought. In many countries a carbon tax could actually be next to neutral (Smith, 1994).

Sectoral distribution

Sectoral consequences have gained less academic attention so far, despite the fact that the political stage has seen fierce lobbying for tax exemptions to more vulnerable industries in both the European Union and the US. Among the few models to provide a sufficient degree of sectoral detail to answer such questions are Jorgenson and Wilcoxen (1992) and McKibbin and Wilcoxen (1992b). Their results imply that, at least for the US, a carbon tax would above all hit the coal industry, and only to a far lesser extent other energy-related sectors such as electric utilities and gas utilities. Few studies exist for countries other than the US (although see Glomsrød et al, 1992; Pezzey, 1992b; Standaert, 1992). On the whole they seem to provide the expected result that energy-intensive industries will be hit hardest. However, ulti-

---

5 For a numerical analysis of these effects see Symons et al (1994).

Table 7.1 The distributional incidence of a carbon tax (Tax payments as % of total household expenditures)

Country	Poorest 25%	Second 25%	Third 25%	Richest 25%	All houses
France					
domestic energy	0.46	0.35	0.28	0.25	0.30
motor fuels	0.15	0.21	0.24	0.24	0.22
total	0.61	0.56	0.52	0.48	0.52
Germany					
domestic energy	0.94	0.79	0.68	0.58	0.69
motor fuels	0.10	0.18	0.21	0.20	0.19
total	1.04	0.97	0.89	0.78	0.88
Ireland					
domestic energy	1.96	1.44	1.06	0.79	1.12
motor fuels	0.13	0.19	0.22	0.22	0.20
total	2.09	1.63	1.28	1.00	1.32
Italy					
domestic energy	0.57	0.49	0.46	0.49	0.49
motor fuels	0.14	0.27	0.28	0.24	0.25
total	0.71	0.76	0.74	0.73	0.74
Netherlands					
domestic energy	0.74	0.59	0.56	0.49	0.56
motor fuels	0.09	0.15	0.15	0.16	0.14
total	0.83	0.73	0.71	0.65	0.71
Spain					
domestic energy	0.43	0.33	0.26	0.26	0.29
motor fuels	0.14	0.20	0.24	0.25	0.23
total	0.57	0.54	0.51	0.51	0.52
UK					
domestic energy	1.89	1.10	0.85	0.62	0.91
motor fuels	0.07	0.18	0.20	0.21	0.19
total	1.96	1.28	1.05	0.84	1.10

Source: Pearson (1992).

mately the distributional incidence will be determined by the outcome of the political battle for tax exemptions.

REVENUE RECYCLING AND DOUBLE DIVIDEND

Any tax likely to have a noticeable impact will raise a considerable amount of revenues. Arguably, undesirable distributional effects should therefore be easy to correct by a corresponding redistribution of tax revenues. However, there appears to be a conflict here between considerations of distributional equity and the idea of efficient revenue recycling. It has been argued in the literature that the imposition of a carbon tax would enable governments to lower other distortionary taxes such as income tax. This would lead to an additional, non-environmental benefit in the form of a reduced deadweight loss from government revenue raising (Pearce, 1991). The size of this so called 'double dividend' could potentially be considerable. Distortionary losses due to taxation are equivalent to 20–50 cents per tax dollar raised, according to Ballard et al (1985).

The exact size of the double dividend depends on the chosen mode of revenue recycling. Different recycling modes, including personal income tax cuts, deficit reduction and investment tax credits, have been analysed by Shackleton et al (1992). They found that the greatest double dividend will be achieved under schemes involving reductions in capital and labour costs which encourage capital formation, and thus long-run growth, rather than immediate consumption (see also Goulder, 1993). A selection of their results is reproduced in Table 7.2. Note that in some cases welfare (excluding greenhouse damage) could actually rise in the long run.

Economic gains from revenue recycling are thus clearly important. Several, more analytical, observations nevertheless arise. First, as has already been pointed out, there is obviously a trade-off here between equity and efficiency. If revenues are used to offset the regressive character of a tax, there will be less scope for a double dividend. Conversely, it appears that maximizing the double dividend effect will aggravate the regressive character

Table 7.2 Costs of a US carbon tax under different recycling scenarios (% change in discounted constant price GNP 1990–2010)

	DRI, Brinner et al (1991)	LINK, Kaufmann et al (1991)	Jorgenson Wilcoxen (1992)	Goulder (1993)
Lump sum tax cuts	−0.58	−0.46	−0.62	−0.24
Deficit reduction	−0.40	−1.02		−0.24
Personal income tax cuts	−0.56	−0.53	−0.16	−0.16
Corporate income tax cuts	0.40	−0.11	0.60	−0.17
Payroll tax cuts				−0.18
– employee only	−0.58	−0.53		
– employer only	0.19	−0.25		
Investment tax credit	1.55	1.67		0.00

Note: Tax scenario involves a $15/tC tax, imposed in 1990, growing at 5 per cent pa.

Source: Shackleton et al (1992).

of a tax (Smith, 1994). The question of the optimal balance is a delicate one, which has not been addressed so far.

Second, if the introduction of a carbon tax has the potential to raise welfare in the long run, this raises doubts about the optimality of the current tax system. If welfare gains can be achieved by shifting the tax burden away from the factors capital and labour, they should be secured independently of environmental considerations. Such optimal taxation questions should, however, not be muddled with the double dividend issue. Double dividends only concern a comparison between the optimal tax system with and without accounting for the externality (Ulph, 1992).

Third, one may wonder whether a double dividend will always occur, whatever the size of tax revenues relative to actual requirements. Ulph (1992) has argued that in a second best world where lump sum transfers are not possible, this will not be the case (see also Pezzey, 1992c). Two problem cases are distinguished. If the revenue requirement is large relative to potential tax revenues, there will be an incentive to overtax pollution, and in the most extreme case the distortion thus created in the environmental market will overshadow the improvements with respect to traditional goods. Conversely, if revenue requirements are relatively small, there is obviously no scope for a double dividend and, in the worst case, losses may even occur if revenues have to be returned through distortionary subsidies.

This raises a fourth question. Arguably, double benefits would similarly arise from other pollution taxes, eg on sulphur, waste or in the transport sector. If the potential for a double dividend is limited and depends on the amount of 'green' tax money raised, which share of the benefits should then be attributed to a carbon tax? The question of double dividends with multiple externalities has so far not been addressed.[6] Taking it up may lead to a more general discussion about the scope and merits of an ecological tax reform, as proposed by von Weizsäcker and Jesinghaus (1992).

---

6 The secondary benefits literature (see Chapter 6) does deal with multiple externalities, but neglects the revenue raising aspect.

Chapter 8

Discounting

DISCOUNTING CONCEPTS

While a general economic and ethical discussion about discounting can be dated back to at least the end of the last century (see eg the references in Markandya and Pearce, 1991), the debate in the context of climate change was launched only recently, initiated mainly by Cline (1992a), and later taken up by eg Birdsall and Steer (1993), Cline (1993b), and Broome (1992). Many studies still bypass the issue, and without much discussion typically use a discount rate compatible with historical savings and interest rate data.[1]

To follow the arguments, it will help to first introduce the different relevant rates of discount. The appropriate discount rate will depend on the choice of the numeraire, ie on the units of measurement to be discounted (see Lind, 1982; Hanley, 1992). The usual distinction is between utility discounting (with well-being or felicity as numeraire) and consumption discounting (with commodities as numeraire).[2] In addition, a distinction has to be drawn between the consumption rate, or social rate of time preference, on the one hand, and the rate of return on capital on the other.

The three concepts – utility discounting, consumption discounting, and return on capital – are connected as follows. If utility is the numeraire, a positive discount rate is used to reflects people's impatience or myopia, ie the fact that utility today is perceived as being better than utility tomorrow. It is also called the pure rate of time preference. As we will see, it is this rate which is at the core of most of the discounting debate.

Moving from utility discounting to consumption discounting, a second effect has to be added to myopia: that of decreasing marginal utility. The incremental utility derived from additional income (or consumption) is generally assumed to be lower, the higher the initial level of income. In the case of utility discounting this is directly taken care of through the shape of the utility function. Here, it has to be added separately.

---

1 Eg Manne and Richels (1992); Peck and Teisberg (1992; 1993a, b); Nordhaus (1992, 1993a, b).

2 In the global warming context, a third possibility has recently emerged, which is the discounting of physical emissions (see Rosebrock, 1993). The approach basically involves the extension of the consumption rate of discount to a (simplified) shadow price of emissions, by incorporating a greenhouse damage function as well as the rate of decay of greenhouse gas concentration.

The consumption rate of discount thus consists of two parts, one accounting for myopia (the pure rate of time preference), and the other for decreasing marginal utility of income. The usual formal notation is[3]

$$\delta = \rho + \omega y \tag{1}$$

where δ is the consumption rate of discount and ρ the pure rate of time preference. ω denotes the income elasticity of marginal utility. It measures by how much the appreciation of additional income varies if the initial income level is changed by 1 per cent. ω has to be multiplied by the rate of growth of per capita income, y, to calculate the total effect.

Note that, although usually assumed to be positive, the effect of decreasing marginal utility on the discount rate may be of either sign, depending on whether future generations will be better off or worse off than today's. Since global environmental problems such as climate change can potentially affect future income, and thus growth rates, the size of the social rate of time preference also depends on the degree of future environmental damage.

Next we turn to the distinction between the social rate of time preference and the rate of return on capital. In a first best world without distortions the two rates would be identical, as is explained in Figure 8.1. In an intertemporal optimum the slope of the indifference curve between consumption today and consumption tomorrow (the social rate of time preference, δ) is equal to the slope of the intertemporal production possibility curve. The production possibility curve PP determines by how much future income can be increased by forgoing consumption today. Its slope is equal to the marginal product of capital, denoted r.

In reality, however, the two rates δ and r will differ, mainly due to the effect of taxation, and the marginal product of capital will be higher than the social rate of time preference. This is easily seen. Suppose the social rate of time preference is δ, ie to be willing to give up consumption of 1 today we need to be compensated with $1 + \delta$ tomorrow. To offer this compensation, an investment project will have to yield a gross return of $r > \delta$, such that after deducting taxes, t, we have $\delta = (1-t)r$.

THE PURE RATE OF TIME PREFERENCE

Having thus set the scene, we can now discuss the main issues of the discounting debate, as they have emerged in the context of global warming. The main controversy is about the pure rate of time preference, in particular whether the notion of impatience, which underlies a positive rate of time preference, is ethically defensible.

There seems little scope to question the significance of impatience at the

---

3 Strictly, the formula only applies to the case of a so-called CRRA (constant relative rate of risk aversion) utility function, see Lind (1982).

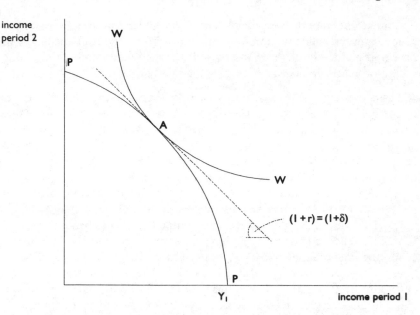

Suppose there are two time periods, I and 2. At the beginning of period I society is provided with an initial income Y_1. Income can be shifted into the future through investment, ie by moving along the intertemporal production possibility frontier PP. The slope of PP is the rate of return on investment, r. By giving up $1 of income in period I, income in period 2 increases by I+r. Successive additional investments yield diminishing returns of future output, so PP is concave. WW is an indifference curve. It shows at what rate society is willing to give up consumption in period I for additional consumption at time 2. The slope of WW is the social rate of time preference. The optimal intertemporal consumption-saving plan is at point A, where the rate of return on capital r equals the social rate of time preference δ. At this point the highest possible welfare level is achieved.

Figure 8.1   Intertemporal choice and the return on capital

level of personal preferences. Empirical evidence from savings and interest rate data suggests a positive pure rate of time preference of about 3 per cent, as reported in Nordhaus (1992).[4] Yet, several authors have criticized the use of a positive rate at the level of society.[5] Ramsey (1928) discarded a positive rate as irrational and ethically indefensible. Pigou (1932), Harrod (1948) and Solow (1974, 1992), among others, found it similarly unpersuasive. Authors who have argued against a positive rate within the global warming context include Cline (1992a, 1993b) and Broome (1992).

The basis for most objections is the impartiality argument, which holds that time should not affect the value of 'good'. In the words of Broome (1992, p 92):

---

4 For a critique of this finding, see Cline (1992a).
5 See eg the discussion in Cline (1992a), Markandya and Pearce (1991), and from a more philosophical angle Broome (1992) and Parfit (1983).

A universal point of view must be impartial about time, and impartiality about time means that no time can count differently from any other. In overall good, judged from a universal point of view, good at one time cannot count differently from good at another.

In terms of intergenerational equity, impartiality implies that the well-being of one generation should not be counted differently from that of any other. Discounting, however, does just this, and places less value on the well-being of future generations. Although the utilitarian view underlying impartiality may itself be subject to criticism, this is probably the most powerful argument against a positive pure rate of time preference.

The question then seems to be whether individual preferences, which clearly exhibit impatience, should have an influence on society's choice of the social rate of discount, where impatience may be questionable. Liberal economic tradition would say yes. It is a fundamental principle of most economic analysis that social decisions should be based on individual preferences (see Markandya and Pearce, 1991). However, in practice counter examples to this principle abound. Drug legislation, safety regulations, speed limits or state pension schemes are all examples of a paternalistic state ignoring individual preferences to enhance overall welfare.

As Markandya and Pearce correctly point out, overruling preferences in this way should require compelling reasons. But then, the list of arguments raised against using a positive pure rate of time preference seems to be rather compelling. If problems of drug abuse and road safety warrant paternalism, it seems hard to find reasons why a fundamental issue like intergenerational equity should not. This then seems to speak for a zero utility rate of discount. In favour of a low rate, it has also been argued that people, although exhibiting a positive rate of time preference as *consumers*, may have different preferences as *citizens*, when considering public investment projects. Another argument suggests that society as a whole may have a higher savings rate than individuals, since saving may cause external benefits (see Hanley, 1992).

DISCOUNTING THE ENVIRONMENT

Conservationists often argue that environmental problems justify a specially adjusted rate of discount. Usually they argue for a low rate since this would give more weight to the long-run benefits associated with environmental goods and would reduce the rate of depletion of natural resources. Providing analytical support for this argument, Weitzman (1994) has shown that in a world with two economic sectors – a polluting one and one engaged in cleaning up – the appropriate rate of discount for the latter sector should be lower than the return on capital in the former. However, calls for a low discount rate may backfire, as has been noted by Markandya and Pearce (1991). They observe that using a lower discount rate will boost investment and lead to the endorsement of additional, potentially damaging projects which would have been considered uneconomic otherwise. A low discount rate may thus also lead to increased environmental degradation, and in some cases the environmental cause would arguably be better served with a high rate. Norgaard and Howarth (1991) have called this ambivalence the conservationist's dilemma.

It would be wrong to think that the conservationist's dilemma is a genuine difficulty of environmental decision making, however. Rather, it appears that the problem mainly arises from the unintended combination of the discounting question with another issue, that of environmental valuation. The dilemma can thus easily be resolved, at least conceptually. It seems clearly wrong to apply unjustifiably high or low discount rates in the hope of thus furthering environmental protection. Concerns about the environmental consequences of a project should not manifest themselves in the discount rate, but as properly valued cost or benefit streams. The socially correct rate of discount can then be determined independently of the potential implications on individual investment projects, eg on the basis of the ethical considerations discussed in the previous section.[6] The point to note then is that as long as the socially correct discount rate is used, whatever its size may be, and as long as a proper project appraisal is carried out which accounts for *all* costs and benefits, no excessive environmental degradation will take place. Degradation will occur only up to the point that is socially justifiable. Of course, in a real world situation, where environmental benefits cannot always be assessed carefully, the point by Markandya and Pearce will again be a valid one, as has been outlined by Broome (1992).

DISCOUNTING AT THE RATE OF RETURN ON CAPITAL

A further point of discussion is whether the rate of return on capital should be preferred to the consumption rate of discount. The use of the rate of return on capital was for example proposed by Birdsall and Steer (1993). They noted that, in a world where investment capital is scarce, it should be invested in those projects which yield the highest return. The correct discount rate should therefore be equal to the rate of return on capital, which represents the opportunity costs of a project. As mentioned above, due to distortions the rate of return on capital is usually higher than the social rate of time preference.

The discounting literature seems to agree that, ultimately, the choice between producer rate and consumer rate depends on the exact character of the costs and benefits of an investment project (see Lind, 1982). Cost or benefit flows concerning consumption should be discounted at the con-

---

6 In the context of the Weitzman (1994) model Pearce and Ulph (1994) have shown that as soon as it is possible to control for externalities through another instrument like an emission tax, a specially low discount rate in the 'clean' economic sector will no longer be justified.

sumer rate, and those related to alternative investment projects at the producer rate. The question is thus about the correct split. Birdsall and Steer argue that all costs and benefits from greenhouse action ultimately affect investment. This view is opposed by Cline (1993b, 1992a), who debits only about 20 per cent of the funds to capital investment, consistent with the average consumption/savings rate.

While it might be true that the costs of abatement may ultimately fully crowd out private investment, as claimed by Birdsall and Steer, their argument seems less compelling with respect to the benefit side, ie with respect to damage avoided. Damages such as the loss of species or an increase in morbidity clearly affect consumption more than investment. At least for the damage side, it seems, the correct rate will therefore have to be a mix between the social rate of time preference and the return on capital.

A case for using the rate of return on capital for consumption-related damages could only be made if funds were set aside and invested at the market rate, in order to provide actual – rather than hypothetical – compensation for future generations (Parfit, 1983). Basically, the future increase in consumption caused by a higher investment today will then have to be sufficient to compensate for future damage, consistent with the idea of sustainable development (see Pearce, 1994a). As was shown by Mäler (1989), the relevant rate would in this case be the future return on capital, rather than today's, and this rate will be lower, the higher the level of intergenerational transfers (Howarth and Norgaard, 1993). The discount rate would become endogenous: the higher future damages, the higher the necessary compensation, and thus the lower the rate of discount. A situation in which market interest rates assure efficiency and transfers are used to obtain intergenerational equity would be first best. However, the argument above clearly depends on the availability of intergenerational transfers. If transfers are not feasible, the most desirable discount rate will in general be different from the market interest rate. In fact, in some cases it may even be negative (Howarth and Norgaard, 1993).

Evidently, the discounting issue is far from settled. Ultimately, the question is a political one which cannot be solved in the academic debate alone. In the absence of intergenerational transfers, there seem to be good arguments in favour of a zero pure rate of time preference as a second best rate for intergenerational projects. However, as long as no definite answer is obtained, the best way forward will probably be for researchers to supplement their results with extensive sensitivity analyses to work out the robustness of their findings.

Chapter 9

The Optimal Policy Response to Global Warming

THE COST-BENEFIT APPROACH

There is hardly an aspect of greenhouse economics which is more fiercely disputed than the question of the optimal policy response. Several approaches to the problem can be distinguished on a methodological level. The most prominent one, at least among economists, is probably the cost-benefit approach. In the cost-benefit approach the optimal policy is determined through a trade-off between the costs of policy action and the benefits from greenhouse damage avoided. This does not necessarily imply a strict cost-benefit analysis in the traditional sense, though, but more generally encompasses the idea that decisions emerge from the weighing up of 'goods' against 'bads'. What the 'goods' and the 'bads' are, and how to weigh them, can again be disputed.

Intertemporal optimization models

Greenhouse gases are typical examples of stock pollutants. It is not the annual *flow* of emissions, but their *accumulation* in the atmosphere which is problematic. Global warming policies are thus best analysed in an intertemporal optimization, or so-called optimal control framework, in which abatement and damage costs are traded-off over several time periods.

The first application of this method to global warming is the influential paper by Nordhaus (1991b, c), although, strictly speaking, the simplified approach used there does not actually constitute a fully fledged optimal control model. Using an aggregate abatement cost function derived from Nordhaus (1991a), and calculating a marginal CO_2 damage of $7.3/tC in the best guess case (see Chapter 4), the paper concludes that only a limited amount of greenhouse abatement would be warranted, most of it achieved through reforestation and the phasing out of CFCs. This controversial conclusion has provoked much criticism.[1] The model has indeed several shortcomings, the most important being its assumption of a resource-steady state and the imposition of a linear relationship between greenhouse damage and emissions (see Chapter 4).

---

1 Most prominently from Cline (1992a), but also from eg Ayres and Walter (1991), Daily et al (1991), and Pachauri and Damodaran (1992).

An alternative to the Nordhaus model has been provided by Cline (1992a). Owing to the inclusion of features like no regrets options and risk aversion, and by using a different discount rate, Cline found favourable benefit-cost ratios for an aggressive abatement plan of freezing CO_2 emissions at 4 GtC per annum, about two-thirds of their 1990 level. His analysis is still not based on an optimal control model, but assesses the merits of one particular project.

Optimal control models developed since provide a similarly differentiated picture. Two main models currently exist, the CETA (Carbon Emission Trajectory Assessment) model by Peck and Teisberg (1992; 1993a, b) and Nordhaus' subsequent attempt, the DICE (Dynamic Integrated Climate Economy) model (Nordhaus, 1992; 1993a, b).[2] Both models are traditional neoclassical growth models, extended by a greenhouse damage sector. DICE is a highly aggregated representation of the world economy with a single representative household producing and consuming one single commodity (income). CETA, on the other hand, also provides a relatively detailed description of the energy sector, based on the Manne and Richels (1992) model.

While Nordhaus' results are very similar to his earlier conclusions – he comes up with optimal CO_2 reductions of only about 15 per cent off baseline projections by 2100 – the results are in general quite heterogeneous and appear to be extremely sensitive to the underlying assumptions. This has been noted by Cline (1992b) who replicated the DICE model and observed that the Nordhaus result of only modest abatement 'does not stem inherently from the optimization model and approach used, but hinges on the particular assumptions applied' (p 31), the most important one being the choice of the discount rate. In some of Cline's alternative calculations CO_2 emissions are virtually phased out by the end of the 21st century. Other critical features include the availability of a carbon-free backstop technology and the treatment of non-market damages. By interpreting non-market damages as a direct element in the utility function, rather than a production cost, Tol (1994) found significantly higher abatement levels than Nordhaus.

A similar, although less extreme, picture also arises from the sensitivity analyses done by Peck and Teisberg (1992, 1993a, b). Over the first 40 or 50 years the various optimal emission trajectories calculated in the model barely differ from each other and all closely follow the baseline. The subsequent trajectory, however, strongly depends on the chosen damage parameters, particularly on the slope of the damage function. Significant emission reductions only occur for a steep damage-temperature relationship (Peck and Teisberg, 1993b).

What then are the conclusions which can be drawn from intertemporal optimization models so far, and what are their principal weaknesses? A disappointing fact about the state of the analysis so far is certainly the high diversity of the results. Although there appears to be a strong tendency in the

---

2 See also Tol (1993) and Maddison (1993).

results to favour more moderate action, it seems similarly true that, at least in the longer term, almost any abatement policy can be justified through the choice of appropriate parameter values. The conclusion following from this fact should be obvious. If the optimal greenhouse policy crucially hinges on the value of certain parameters it seems natural to intensify the efforts toward a better understanding of them. On the scientific side, this means that more effort will have to be put into the analysis of long-term climate effects to determine the slope of the damage function. The search also has to start for the identification of climate thresholds. On the economic side, questions of discounting and of intergenerational equity in general will have to receive more prominence, specially in the light of a general critique on cost-benefit analysis in this respect.[3]

Integrated assessment

Most of the discussion so far has revolved around the question of CO_2 abatement. However, available options are wider than that. A more comprehensive view should therefore be adopted which also includes geo-engineering options, carbon removal and sequestration, the abatement of other greenhouse gases as well as the various adaptation options. A comprehensive approach would also take into account the spill-over effects to other policy areas, eg in the form of secondary benefits or as a double dividend (see Chapters 6 and 7). The problem is to find the optimal level of each of these activities. None of the optimal control models is comprehensive enough to do this.

The analysis is usually confined to the abatement of at most one or two gases other than CO_2. Of the non-energy related abatement options carbon sequestration has gained some prominence.[4] Few authors have studied the scope for carbon removal techniques (eg Okken et al, 1991). Geo-engineering has gained little prominence in the economic literature so far, and, on the whole, does not seem to constitute a serious alternative to greenhouse gas abatement at the moment.[5] Adaptation is typically only included implicitly, in that most studies assume a damage function based on a cost-efficient adaptation policy (see Chapters 2 and 5). What exactly this optimal response is, however, is usually not specified. Nor are regional differences in climate vulnerability and adaptation levels normally taken into account.

The need for a more comprehensive approach has been recognized and a series of so called *integrated assessment models* are currently being built, which aim at providing a more detailed picture of the link between economic activities and climate change (see Figure 9.1).

---

3 See Howarth and Norgaard (1993), Howarth and Monahan (1992). See also Chapter 8.
4 See for example Sedjo and Solomon (1989), Dudek and LeBlanc (1990), and Cline (1992a).
5 Several geo-engineering options, such as the emission of radiation absorbing particles into the stratosphere to artificially cool the earth, are listed in Nordhaus (1991a). They are however not considered further.

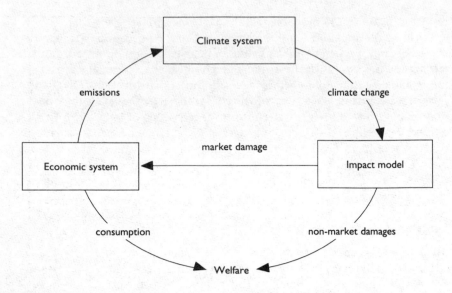

Figure 9.1   Integrated assessment of climate change

Combining descriptions of the socio-economic activities causing emissions with representations of the bio-geochemical processes leading to climate change and climate change impact models is a formidable task, and most models are (or will be) rather demanding in terms of computation time. Due to their complexity, many models do not (yet) 'close the loop', and climate damages do not always feed back to the economy. Contrary to the optimal control models discussed above – which are, of course, integrated assessment models as well, although on a much smaller scale – they do therefore, usually not allow to directly calculate an optimal emission path. A short description of six major integrated assessment projects is given in Box 9.1.

HEDGING AND UNCERTAINTY

It is one of the main weaknesses of most greenhouse cost-benefit models that they tackle the problem as if all parameters were known with certainty.[6] Clearly, this is not the case, and an optimal greenhouse policy has to take uncertainty into account. In the terminology of Chapter 2, decision-makers should be interested in the entire damage probability distribution, and not just its mode (see Figure 2.2). Incorporating uncertainty raises a host of additional issues dealing with risk aversion, irreversibilities and the possibility of climate surprises. It also raises the question about the optimal mix

---

6 Exceptions are Hope et al (1993) and Dowlatabadi and Morgan (1993).

Box 9.1
Main integrated assessment projects

MERGE (Manne et al, 1993)
The Model for Evaluating Regional and Global Effects of greenhouse gas reduction policies is an extension of the Manne and Richels (1992) energy model (see Box 6.1), expanded to the year 2200 and combined with comparatively simple climate and damage representations.

IMAGE 2.0 (Alcamo et al, 1993)
Probably the first integrated assessment project, the Dutch Integrated Model for the Assessment of the Greenhouse Effect concentrates on the connection between emissions, climate change and climate impacts, and incorporates less economic detail. Version 2.0 of the model consists of three fully linked sub-systems: Energy-Industry, Atmosphere-Ocean and Terrestrial Environment.

PAGE (Hope et al, 1993)
The main feature of the PAGE (Policy Assessment of the Greenhouse Effect) model is its treatment of uncertainty. All major parameters are treated as random. As a consequence of this emphasis on uncertainty analysis the model is kept relatively small, to facilitate repeated runs with different parameter values.

ICAM (Dowlatabadi and Morgan, 1993)
The Integrated Climate Assessment Model developed at Carnegie Mellon University is also a comparatively small model. As with PAGE, a lean structure is necessary to incorporate uncertainty analysis, one of the main features of the model.

GCAM (Edmonds et al, 1993)
The Global Change Assessment Model combines the Edmonds et al (1992) Second Generation Model (see Box 6.1) with impact components mainly from the MINK study (Rosenberg and Crosson, 1991) and two atmospheric models, one for atmospheric chemistry and one for climate dynamics.

MIT
Probably the computationally most demanding endeavour, the project of the Massachusetts Institute of Technology will combine the GREEN model (Burniaux et al, 1992b, see Box 6.1) with a small-scale version of the GISS Global Circulation Model and a terrestrial ecosystems model.

---

between greenhouse gas abatement and further research to resolve uncertainty.

Greenhouse gas abatement as insurance

Many advocates of immediate action interpret greenhouse gas abatement as an insurance, and argue that for the same reasons as individuals are willing to secure themselves against hazards of all sorts, society should be willing to spend some money to protect itself against adverse climatic effects, particularly since these may be irreversible and potentially disastrous. This view is implicit in much of the literature about the precautionary principle (see eg O'Riordan, 1992), and it has also been issued by eg Schneider (1993), Schelling (1992) and Cline (1992a) who in his analysis accounts for risk aversion by assigning additional weight to the worst case scenario. It is also consistent with the quasi-option value literature (eg Arrow and Fisher, 1974), which shows that over-protection is efficient in the presence of irreversible damage.

Explicit analyses of the insurance argument include Manne and Richels (1992, 1991) and Parry (1993). A related analysis is that by Krause et al (1989). Manne and Richels (1992, 1991) use their Global 2100 model to illustrate how in a situation where governments have to act before uncertainties are resolved, they should hedge against a possible future carbon constraint by doing more abatement than they otherwise would. In an other paper they further argue that agents will be hedging in a similar way, so that the mere possibility of future action acts as an implicit carbon tax. A 50 per cent chance of an emission stabilization policy being in place from 2010 onwards, for example, would affect emissions in the same way as a $17.50/tC carbon tax under certainty (Manne and Richels, 1994b).

An estimation of the value of reduced climate risk has been carried out by Parry (1993), who somewhat surprisingly found a rather low value for a reduction in climate risk. The main reason for this seems to be discounting and the fact that most of the damage will occur only late in the future. It should, however, also be noted that Parry's analysis is based on the not fully satisfactory Nordhaus (1991b, c) model, and that he rather optimistically assumes a $2xCO_2$ damage distribution which is symmetric around a mean of zero.

A completely different stance towards the uncertainty issue has been taken by, among others, Nordhaus (1993c), who argues that, rather than implying precautionary abatement, uncertainty should be a reason to postpone action until better information is available. It would be irresponsible, so the argument goes, to commit resources to a problem the severity of which is still uncertain. Or, to put it another way, greenhouse gas abatement is too risky a project to invest in. Lempert et al (1994) have similarly argued that the costs of delaying carbon abatement for a decade would only marginally affect the costs of potentially having to achieve a stringent target later on. Swift action should consequently only take place if the probability of a tight emissions policy in the future is sufficiently high, and the expected policy measures are sufficiently severe. However, Broome (1992) has rebutted wait-and-see arguments of this type by observing that it is not the riskiness of an individual project which matters, but the overall risk that society is exposed to, which he argues may well be reduced through early greenhouse gas abatement.

The academic controversy notwithstanding, the Climate Convention has endorsed the insurance argument. Article 3 of the Convention recognizes the need for precautionary action in order to avoid serious or irreversible damage (see Grubb et al, 1993).

Low probability/high impact events

Much of the persuasiveness of the insurance argument hinges on the fear of a possible climate catastrophe.[7] The question of how to handle such high impact/low probability events is, however, far from settled and deserves much further attention. Two archetypal methods of decision theory have been applied so far. The first is the *expected utility* approach, as used by eg Heal (1984) and Peck and Teisberg (1994). The second is a *maximin* approach, as proposed by eg Krause et al (1989).

In Heal's model the atmosphere is interpreted as an exhaustible resource of unknown stock. Exhausting the resource (passing a climate threshold) would trigger a climate catastrophe, and utility would fall to zero. People take precautions against this event by reducing emissions, and the optimal rate of emissions is determined in a trade-off between abatement costs and the risk of catastrophe. The amount of abatement undertaken crucially depends on the degree of people's risk aversion, as well as on their perception of the likelihood of a catastrophe. Peck and Teisberg (1994) have implemented Heal's idea using the probability assessments of the Nordhaus (1993d, 1994) poll of experts (see Table 2.3 above). Surprisingly, their results hardly differ from those achieved with a conventional damage function (eg Peck and Teisberg, 1992). In the main scenario – with catastrophe probabilities calculated using data from all respondents – CO_2 emissions continue to rise throughout the next century and reach almost 40 GtC/year by the year 2100, about six times 1990 levels. In a more pessimistic scenario based on the responses from natural scientists only, emissions still rise to about 20 GtC/year in 2100, and are only gradually reduced below 1990 levels thereafter.

With a maximin approach the picture is completely different. Under a maximin strategy society is only concerned with the worst possible outcome (ie an early catastrophe) and implements the policy which would maximize the payoff under this scenario. As a consequence the resulting abatement targets are rather high. In the analysis of Krause et al (1989) it leads to the imposition of a maximum warming target of 0.1°C/decade, and 2.5°C in total. Working backwards they found that this would translate into a reduction in emissions of 20 per cent below 1985 levels by 2015 and 75 per cent by 2050.

Neither of these approaches is fully satisfactory. The predominant paradigm – expected utility theory – is often at odds with observed human behaviour, which may affect its relevance for the analysis of low probability/

---

7 Examples of potentially catastrophic climate change impacts have been given in Chapter 2.

high impact events (see eg Machina, 1982). In addition, the method crucially depends on the knowledge of a probability distribution for the occurrence of a climate catastrophe. While methods exist to attain subjective probabilities, reliable figures may be difficult to obtain. The maximin approach, on the other hand, appears to be somewhat extreme in its radical risk aversion, and its negligence of both the probability of occurrence of a catastrophe (in as far as known) and the costs of achieving the target are somewhat irritating. Individuals are willing to take risks, if the reward from doing so is sufficiently high and the probability of an accident sufficiently low. Although the situation of individuals may be different from that of society as a whole, this fact should not be ignored.

Irreversibility and learning

Kolstad (1991) has analysed the difference between the 'wait and see' and the 'insurance' school in the context of irreversibilities and learning. He concluded that the result from the quasi-option value literature – under-investment in projects with irreversible consequences – will generally hold, but argues that there may be irreversible consequences on both the abatement and the damage side. Global warming may cause irreversible damage, eg to ecosystems, but in a similar way greenhouse gas abatement may involve investment in special abatement technologies which afterwards become sunk costs. Depending on which side dominates, uncertainty can then justify both excessive and moderate abatement.

In subsequent simulations with a variant of the DICE model Kolstad (1993) found evidence that, should investment costs indeed be sunk, the capital cost element may dominate in the case of global warming, although he raises the questions of whether the irreversible investment assumption is indeed a reasonable one.[8] Damage irreversibilities, on the other hand, have only a limited impact, and the reason for this is again the long time delay until their occurrence. As with other optimal control results (see previous section, 'The cost-benefit approach'), this outcome may thus be an artefact of the chosen parameter values, in particular the discount rate.[9]

---

8 The results by Manne and Richels (1991, 1992, 1994b), which are based on a model with a very detailed energy sector, suggest that there are little sunk costs. Their optimal emission path, which exhibits hedging behaviour as long as there is uncertainty, very rapidly swings back to the perfect information trajectory once uncertainty is resolved. The opposite view is taken by Hourcade (1993), who emphasizes the importance of bifurcations, long run structural decisions after which countries are locked in to a certain energy path.

9 The Kolstad result has also been questioned on more theoretical grounds by Ulph and Ulph (1994a). They argue that the quasi-option value result (eg Arrow and Fisher, 1974) does not necessarily extend to stock pollutants like CO_2. They show that the usual result will only hold if the irreversibility constraint is actually binding in the case without learning. In other words, the usual result only holds if, in a situation where uncertainty is never resolved, optimizing agents would in principle prefer to cause some damage in earlier periods and undo it later, but are prevented from doing so because impacts are irreversible.

The value of information

Kolstad has analysed the case of autonomous learning, in which further information is acquired automatically over time as more observations become available. The more interesting question of how far the resolution of uncertainty should be accelerated through increased research has so far hardly been addressed. A first step in this direction involves an analysis of the value of increased information, and such calculations have been carried out by Peck and Teisberg (1993b) and Manne and Richels (1992, 1991).

The (expected) value of information can be defined as the difference between the payoff in the case of uncertainty, when a hedging strategy is followed, and the expected value of the payoff obtained under certainty. The value of information differs depending on the reliability of the received information, on the time it becomes available and on the uncertainty aspect it concerns. According to Peck and Teisberg, the value of a better under-standing of greenhouse damage costs is almost the same as that of an increased scientific knowledge about the extent of warming, see Table 9.1. With respect to damage costs, information about the slope of the damage function is more valuable than information about the consequences of $2xCO_2$. This would suggest that more research should be put into the analysis of the long-term impacts of global warming, as has already been proposed in 'The cost-benefit approach', (see also Chapter 2).

Table 9.1   The economic benefit from acquiring information (Difference in Present Value Payoff, bn$)

Parameter	Information now vs never	Information now vs in 20 years
Climate sensitivity (Warming under $2xCO_2$)	148.0	6.0
Damage under $2xCO_2$	27.0	0.2
Slope of damage function	98.0	0.0
Total	273.0	6.2

Source: Peck and Teisberg (1993b)

THE CARBON BUDGET APPROACH

An alternative to the cost benefit view is the *cost efficiency* or carbon budget approach. Rather than determining the optimal abatement level as a trade-off between costs and benefits it is exogenously determined according to political, scientific or ethical considerations. The aim is then merely to find the cost-efficient way of achieving the target. Studies based on a carbon budget approach include, for example, Anderson and Williams (1993) and Richels and Edmonds (1993). The method has particularly been endorsed by scientists and the political community. Targets have been proposed in several contexts and by several bodies: the Toronto target of a 20 per cent emission cut, for example, the Rio target of emission stabilization at 1990 levels, or the scientific targets set by the IPCC (see Chapter 1).

There seem to be two different, but connected, arguments leading to the endorsement of the carbon budget approach. The first one is based on uncertainty and calls for the implementation of carbon targets as part of a risk minimization policy. This is the maximin strategy against a potential climate catastrophe discussed in the previous section. It is in the tradition of the literature on safe minimum standards (Bishop, 1978). Proponents of such an approach include Swart and Vellinga (1994), Krause et al (1989) and Wirth and Lashof (1991). To minimize the risk of a climate catastrophe the approach requires that an emission target be set at the maximum level of emissions under which a climate catastrophe can reasonably be excluded (see previous section).

The second argument relates to the monetisation of global warming impacts. It questions whether the impacts of global warming can at all be expressed in monetary terms, and indeed whether doing so is reasonable in the first place. The absence of damage estimates then implies that abatement targets have to be determined on different grounds, according to political, social or ethical considerations. Without explicitly endorsing the carbon budget approach, doubts about the quantification of greenhouse damage have, for example, been raised by Broome (1992), who would prefer to see global warming impacts as a change in the course of history rather than a reduction in GNP. Howarth and Monahan (1992) point out that global warming impacts are probably too complex to be assessed even in physical terms, which renders a cost-benefit analysis impossible.

The question which immediately occurs is how carbon targets should, alternatively, be set. Broome's endorsement of the teleological approach raises the suspicion that this would again involve a cost-benefit trade-off of some sort, this time simply based on an implicit assessment of greenhouse consequences, rather than an explicit one. Indeed, it can reasonably be argued that, for example, the initial reluctance of many governments to endorse the Rio target of stabilizing greenhouse gas emissions at 1990 levels is a sign that implicitly the benefits of such a measure were deemed lower than the costs (Pearce and Fankhauser, 1993).

Alternatively targets could emerge from ethical obligations, ie they could be warranted for moral reasons, in the same way as, for example the biblical 'thou shalt not kill' stands independently of possible cost-benefit considerations. This seems to be the stance taken by Howarth and Monahan (1992), who propose a rule for greenhouse action based on the *sustainability principle*. However, while traditional cost-benefit analysis indeed often neglects equity issues, their inclusion may not necessarily make a trade-off between 'goods' and 'bads' redundant. Howarth and Monahan themselves implicitly acknowledge this, when proposing greenhouse action in favour of future generations only 'if doing so would not noticeably diminish ... [today's] quality of life' (pp 6–85), that is, if the *costs* born by today's generation are sufficiently low, presumably relative to *future benefits*. The merits of the rights approach have also been discussed by Broome (1992), who dismisses it in favour of the teleological approach.

CONCLUSIONS

A majority of economists today seem to have endorsed cost-benefit analysis, in the broader sense of the definition given at the beginning of the chapter, as their preferred normative approach (Howarth and Monahan, 1992). Yet, most of them would probably also agree that no single method is capable of providing all the answers. Equity and precautionary arguments will therefore have to complement the more efficiency-oriented cost-benefit considerations. However, while cost-benefit analysis is a well-understood and clearly defined concept, operational definitions for these complementary notions have yet to emerge.

The variability and high sensitivity of available results makes it difficult to derive policy conclusions. Existing optimal control models, on the one hand, have the tendency to favour relatively moderate abatement levels. A risk minimization strategy, on the other hand, would require significant emission cuts within the next two decades or so. Fortunately, the problem is less severe once it is recognized that decisions can be taken sequentially (Manne and Richels, 1992). While model predictions heavily differ over the medium and long term, there is far less divergence with respect to the immediate future – the time period with which current abatement decisions are concerned. Cline (1993a, p 28) even observes 'a surprising convergence of the various analyses' for the first decade, with a 10 per cent to 15 per cent emission cut emerging as a possible consensus policy for the next ten years or so. For subsequent periods, there are signs that the inclusion of secondary benefits and revenue recycling into cost-benefit analyses could tilt the balance in favour of a more stringent abatement policy in the medium run, consistent with the precautionary views held by many authors. This should leave at least some time to learn more about the emissions policy required in the long term. Insuring the world against global warming may not be too expensive, after all.

Chapter 10

International Cooperation

SHARING THE ABATEMENT BURDEN

Efficiency

Greenhouse gas abatement should be carried out where it is cheapest to do. In addition to cost efficiency in the domestic context, which has largely been the topic of Chapter 7, a least cost solution should also be sought at an international level. If domestic efficiency involves equalizing the marginal costs of abatement between the different sectors of an economy, international efficiency requires their equalization between countries. It is well known that most of the proposals discussed in the policy debate so far – such as uniform percentage reductions or emission stabilization across countries – will generally not lead to this outcome. Uniform schemes may cause countries facing high abatement costs to undertake abatement which could more cheaply be achieved elsewhere (see Hoel, 1991a). Several numerical models have produced empirical illustrations of this fact. Table 10.1 shows estimates of the cost difference between a uniform reduction agreement and efficient cooperation, achieved through a uniform carbon tax.

Table 10.1   Efficiency gains from emissions trading (2 percentage point reduction scenario)

		Edmonds Reilly		GREEN		Manne Richels	
		Tax ($/tC)	GDP loss (%)	Tax ($/tC)	GDP loss (%)	Tax ($/tC)	welfare loss
2020	equal cut	283	1.9	149	1.9	325	–
	equal tax	238	1.6	106	1.0	308	–
2050	equal cut	680	3.7	230	2.6	448	–
	equal tax	498	3.3	182	1.9	374	–
2100	equal cut	1,304[a]	5.7[a]	–	–	242	8.0[b]
	equal tax	919[a]	5.1[a]	–	–	208	7.5[b]

[a] year 2095
[b] discounted consumption losses through 2100, trillions of $

Source: OECD (1993).

Just how the economically efficient solution would look is, however, unclear. The question has been analysed, for example by Martin et al (1992) using their GREEN model, and Kverndokk (1993), using a reduced form abatement equation based on Manne and Richels (1992). A cost-efficiency scenario – achieved through a global carbon tax – was also part of the OECD model comparison (see OECD, 1993). The results are rather ambiguous. GREEN and the Edmonds and Reilly model predict that, compared to equiproportionate cuts, a cost-efficient scheme would shift the abatement burden away from rich and semi-rich countries to the developing world.[1] Manne and Richels, and thus also Kverndokk, come to exactly the opposite conclusion. The difference seems to hinge on the treatment of coal. While Martin et al make out cheap fuel switching opportunities in coal-based economies such as China, Kverndokk argues that for a poor country like China switching from cheap domestic coal to imported oil and gas will be relatively expensive.

Equity

The existence of regional disparities raises the question of side-payments. Much as in the domestic case, the above efficiency criterion has to be complemented by equity considerations. In the international context, the question is mainly about a just distribution of the abatement burden between the countries of the developed and the developing world. Analytically, distributional issues are often examined in the context of the allocation of tradeable emission permits. This is mainly for analytical convenience, though, as it allows for a straightforward separation between the equity and efficiency issues.[2] The question could, of course, equally be treated in terms of the redistribution of tax revenues, for example, or proceeds from a permit auction, or indeed the size of transfer payments in general.[3]

A variety of allocation criteria have been discussed in the literature, including the distribution of emission rights in proportion to welfare, population, land area, previous greenhouse gas emissions, as well as several variants and combinations of these.[4] The choice between them is influenced by various criteria of social justice on the one hand, and by the necessities of *realpolitik* on the other (Grubb et al, 1992). That is, the outcome should be

---

1 Detailed descriptions of GREEN and the Edmonds and Reilly model can be found in Burniaux et al (1992b) and Edmonds and Barns (1991), respectively. See also Chapter 6.

2 Note, however, that because of the public good character of global warming, the traditional separation between efficiency and equity will not always hold (Chichilnisky et al 1993).

3 There are schemes which for equity reasons foresee different abatement targets for different groups of countries, such as stabilization of emissions in developed countries only. Note however that these, much like uniform reduction schemes, are unlikely to fulfil the efficiency criterion.

4 See eg Rose (1992), Barrett (1992a), Kverndokk (1993), Grubb et al (1992), or Grubb (1989).

both fair and feasible in the sense that countries are willing to subscribe to it. The aspect of feasibility has particularly been emphasized by Barrett (1992a, b) and Hoel (1991a), who maintain that a treaty will only be signed if it is beneficial to a sufficient number of participants (see following section, below).[5]

As opposed to the case of efficiency, there is no universally agreed definition of equity. Rose (1992; 1990) has put forward several ethical criteria to determine what 'fairness' may mean, and they are summarized in Table 10.2. The table includes, among others, horizontal equity (encompassing egalitarianism as its most extreme form), vertical equity (encompassing Rawlsian maximin), ability to pay, but also the Pareto rule and consensus-building, which opens the door to the *realpolitik* view of Barrett and Hoel. Although emphasizing the importance of consensus building, ability to pay and vertical equity – and thus putting the ball into the corner of developed countries – Rose is rather reluctant to identify a single preferred criterion. Other authors are less so. Grubb (1989) and Kverndokk (1992) advocate an allocation according to population, pegged to a base period or restricted to adult population to avoid perverse incentives on population growth. Barrett (1992a) has put forward a 'Kantian' rule, according to which countries have to abate at least as much as they would wish the others to do. Grubb et al (1992) see a mixed formula as the only feasible option.

One view of the developing world has been made explicit in Agarwal and Narain (1991), who detect 'environmental colonialism' in the debate, and who emphasize the importance of historical responsibilities. They further criticise the greenhouse gas emission statistics of the World Resources Institute (see WRI, 1990) and propose that emission accounts should be net of the carbon sinks found in a country. The lively discussion provoked by their contribution shows that the question of burden sharing is far from settled, and that the discussion has probably only just started. Inevitably the issue will have to be decided in the context of the actual negotiations towards a Climate Change Protocol.

INTERNATIONAL COOPERATION AND THE NEED FOR SIDE-PAYMENTS: AN ILLUSTRATION

At the end of the day the outcome of the negotiation process will perhaps not so much depend on the question of fairness discussed above, as on the acceptability and feasibility of a proposal. Even though Rose (1992) claims that countries would be willing to accept an outcome which is to their disadvantage, as long as it is fair, the more down to earth stand taken by

---

5 Even if this is achieved, there may still be a problem about stability. That is, even if a certain agreement is generally preferred to a situation without, there is no guarantee that countries will not find it even more profitable not to join and to free-ride on the efforts of others. See 'Unilateral action', below.

Table 10.2 Equity criteria for global warming policy

Criterion	General description	Distribution of permits: operational rule	Distribution of permits: reference base[a]
Horizontal	Persons in the same group are treated equally.	Equalize net welfare change (net cost of abatement as proportion of GNP) across nations.[b]	GNP, (land area, energy reserves, CO_2 emissions).
Vertical	Greater concern for the disadvantaged.	Progressively distribute permits (net costs inversely correlated with per capita GNP[b]).	GNP, (land area, energy use, energy reserves, CO_2 emissions).
Ability to pay	Parties pay according to their means.	Equalize abatement costs (gross costs as a proportion of GNP[b]) across nations.	GNP.
Sovereignty	Each nation/person is guaranteed a minimum of basic rights and resources.	Cut back emissions proportionally across nations.	CO_2 emissions.
Egalitarian	Treat every human being equally.	Cut back emissions in proportion to population.	Population.
Market justice	Free market is a fair means of allocation and distribution.	Auction entitlements to highest bidder.	–
Consensus	A decision is fair if parties agree to it.	Distribute permits so that majority of nations is satisfied.	(Population.)
Compensation	Pareto rule: no party should be made worse off.	Distribute permits so that no nation suffers a net loss of welfare.	GNP, (energy reserves).
Rawl's maximin	Maximize welfare of the worst off nations.	Distribute large proportions of permits to poorest nations.	GNP.
Environmental	Emphasizes primacy and 'rights' of ecosystems.	Cut back emissions to maximize environmental values.	CO_2 emissions, (energy use, land area).

a Parentheses indicate applicability is weak
b Net costs: abatement costs – abatement benefits + permit purchases – permit sales. Gross costs: abatement costs only.

Source: Rose (1992).

Barrett (1992a, b) and Hoel (1991a) is perhaps more realistic. They argue that in the absence of an international legislative body able to enforce an agreement, a treaty will only be signed if it is beneficial to all participants. The main distributional problem is thus not equity as such but finding an allocation rule which guarantees a beneficial outcome for a sufficient number of countries.

The global warming game

Whether or not a certain agreement is beneficial to a country depends on several factors. What is basically needed is a comparison between the abatement cost and benefit streams with and without the agreement. Several game theoretic papers have examined this problem, albeit in a rather stylized way.[6] Global warming is analysed as an open access resource problem. The atmosphere is used jointly by all countries of the world to dispose of, among other substances, CO_2, which is emitted as a by-product of GNP-producing activities. However, the accumulation of CO_2 in the atmosphere has negative effects in the form of future climate change, and this not only for the emitting country but for the world as a whole. That is, each emitter imposes a negative externality on the rest of the world.

In the absence of an international agreement this externality is not taken into account. Countries are only concerned with the global warming damage affecting themselves, and their optimal level of abatement will be achieved at the intersection of the marginal abatement cost curve with the individual marginal benefit curve of damage avoided in the country itself. The situation is depicted in Figure 10.1 and the optimal abatement level in the absence of cooperation is a_N. Note that even in the absence of international cooperation a certain degree of abatement will be justified, on the basis of expected individual damages. Countries do not remain in a 'do nothing' position.

The non-cooperative outcome is compared to a cooperative solution in which the externality is fully taken into account. The optimum in the cooperative case is achieved at the point a_C, where the marginal abatement costs of a country equal the sum of all marginal damages world-wide. Point a_C also constitutes a social optimum. However, while desirable for the world as a whole, the socially optimal outcome a_C may not necessarily be beneficial for each individual signatory. It may be that countries with relatively cheap abatement opportunities will be asked to explore them to the benefit of others. These countries will initially be worse off and it will be necessary to compensate them through resource transfers.

---

6 See Fankhauser and Kverndokk (1992), Eyckmans et al (1993), and Tahvonen (1993). Ulph and Ulph (1994b) have analysed the case of international cooperation in the case of uncertainty.

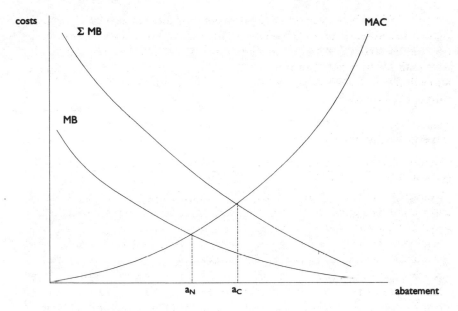

MAC$_i$ marginal abatement costs of country i
MB$_i$ marginal benefits to country i (damage avoided in country i)
ΣMB$_i$ marginal benefits world-wide (damage avoided world-wide)

Figure 10.1   Cooperative vs non-cooperative equilibrium

Some illustrative results

Table 10.3 summarizes the results of illustrative simulations of an international CO$_2$ reduction agreement for the year 2000. The table shows
estimates for the equilibria a$_N$ and a$_C$, the two outcomes suggested by
economic theory, and contrasts them to the political outcome so far, that is
the UN Framework Convention on Climate Change (Table 10.4). Figures
are taken from Fankhauser and Kverndokk (1992), a static model which
combines an abatement cost function based on Manne and Richels (1992)
with a simplified version of the damage model of Appendix 1. It should go
without saying that, given the great uncertainty prevailing particularly with
respect to greenhouse impacts, and the simple character of the underlying
model, the figures are indicative only. They nevertheless offer some general
insights into the economic incentives for greenhouse gas abatement in different geopolitical regions.

The non-cooperative case
In the absence of an international agreement the model predicts that
abatement efforts will probably be restricted to OECD nations other than
the US, with emission reductions of 2 per cent to 11 per cent compared to
business as usual (BAU). For the non-OECD countries (ex-USSR, China

Table 10.3 Optimal abatement levels and welfare gains

Non Cooperative Case (% emission cut relative to business as usual)

	USA	Other OECD	Ex-USSR	China	Rest of world	Total
Low damage	0.9	2.0	0.0	0.0	0.1	0.7
Medium damage	2.2	4.6	0.1	0.1	0.5	1.7
High damage	5.4	10.7	0.2	0.1	1.8	4.3

Cooperative Case (% emission cut relative to business as usual)

	USA	Other OECD	Ex-USSR	China	Rest of world	Total
Low damage	3.2	3.9	0.7	0.6	0.8	2.0
Medium damage	7.9	9.2	2.5	2.4	2.6	5.4
High damage	19.7	21.7	9.3	9.2	9.2	14.6

Welfare Gains from Cooperation (% of GNP)

	USA	Other OECD	Ex-USSR	China	Rest of world	Total
Low damage	−0.000	0.003	−0.000	0.001	0.008	0.003
Medium damage	0.001	0.023	−0.010	−0.000	0.051	0.017
High damage	0.055	0.221	−0.155	−0.094	0.356	0.143

Source: Fankhauser and Kverndokk (1992), revised.

Table 10.4 Abatement commitments under the Climate Convention (% emission reduction in the year 2000, compared to business as usual)

USA	Other OECD	Ex-USSR	China	Rest of world	Total
13	16	0–11[a]	0	0[b]	7–9[a]

a The stabilization call in principle includes Central and Eastern Europe and the former USSR, but these countries are allowed 'a certain degree of flexibility' (see Grubb et al, 1993)
b Neglecting Central and Eastern Europe

Source: Own calculations, after Manne and Richels (1992).

and the rest of the world), the changes in emissions are almost negligible in all scenarios. This may be surprising at first sight, since greenhouse damage in developing nations is expected to be substantially higher than world average, relative to GNP (see Chapter 3). However, as Figure 10.1 reminds us, the estimates result from the simultaneous consideration of both damage and abatement costs. The costs of CO_2 abatement are in general also higher in developing countries, compared to the industrialized world, and this effect dominates over the damage side.

It is interesting to contrast the non-cooperative emission levels with the Rio call for emission stabilization in developed countries at 1990 levels.[7] The estimated abatement requirements, derived from the Manne and Richels (1992) emission scenarios, are reproduced in Table 10.4. A comparison of Tables 10.3 and 10.4 shows that a considerable share, in some cases almost half, of the abatement required under the Rio target could be in countries' self-interest, and would thus be undertaken anyway. It would then be no surprise that most OECD countries have committed themselves to the stabilization target, although in some cases rather reluctantly.

The social optimum
The optimal reduction rates estimated in the model vary quite a lot depending on the different damage cost scenarios, the highest cut being about 15 per cent, compared to the BAU baseline (see Table 10.3). The figures are thus roughly within the range of other cost-benefit results (see Chapter 9). They can again be compared to the Rio targets of Table 10.4. Emission stabilization is roughly consistent with the calculated social optimum for the two OECD regions and the world as a whole, except in the low damage scenario. This may cause surprise, given the perceived modesty of the Rio target. A caveat is however required. Recall the discussion of the cost-benefit approach to global warming in Chapter 9. The model presented here is a typical example of this genre, and therefore also shares its shortcomings. In particular, the model is deterministic, neglecting questions of risk aversion and uncertainty. In addition, the analysis is limited to the year 2000. As argued in Chapter 9, further abatement after the year 2000 may be justified for precautionary reasons, and once spill-overs (secondary benefits, double dividend) are included into the analysis. The Convention itself explicitly states that precautionary measures may be required '[w]here there are threats of serious or irreversible damage' (Article 3, see Grubb et al, 1993).

Side Payments
Table 10.3 also shows the welfare gains from cooperation, that is the difference between the net benefits in the cooperative and non-cooperative equilibria. With less than 0.15 per cent of gross world product, the overall

---

7 Note that according to the Climate Convention stabilization extends to all greenhouse gases. Here we only consider CO_2 from fossil fuel combustion.

welfare gains from cooperating seem rather modest, although at $7–45 bn (for the two high damage scenarios) they are quite substantial in absolute terms. It is, however, likely that the model underestimates the true gains from cooperation, since it assumes cost efficiency within each of the five regions. That is, there is a high level of cooperation (eg between the countries of the former Soviet Union) already in the non-cooperative case.

More important than the absolute values are the directions of the welfare gains, which can give an indication about a region's willingness to cooperate. The highest gains occur in the 'other OECD' countries and the rest of the world. These are also the only regions for which the welfare gains are positive in all scenarios. For the former Soviet Union the gains are mostly negative, while for China and the US the picture is mixed, both regions being more or less equally off, except for the most pessimistic damage case. The calculations would then imply that only OECD countries, with the possible exception of the USA, have an incentive to sign an international agreement. China and the former USSR will be reluctant to do so without side-payments. Contrary to common belief the model would see other OECD countries as the main paymaster, rather than the US. As for the rest of the world, it is important to remember that it is merely a residual category, which consists of regions as different as Eastern Europe, OPEC, Asia and Africa. Hence it is not possible to outline an overall uniform strategy which is optimal for all these countries.

The figures are highly sensitive. Somewhat different results to those reported here have, for example, been obtained by Tahvonen (1993), who sees China as the main beneficiary from an agreement, the main losers being the former USSR, and, to a larger extent than here, also the US. His results are, however, dominated by the effects of later abatement efforts and are therefore not necessarily comparable to the picture for the year 2000 drawn here. Incentive structures may also change once uncertainty is taken into account, as was shown by Ulph and Ulph (1994b).

UNILATERAL ACTION

The free rider problem

Despite the fact that the UN Framework Convention on Climate Change has been signed by over 160 countries, the assumption of full cooperation in a Global Warming Protocol is somewhat unrealistic. This conclusion can be derived from the experience of past negotiations and the attitude of many governments so far. As a matter of fact, even in the Convention itself the emphasis is on action from only a subset of countries, those of the industrialized world (the so called Annex 1 countries, see Chapter 1).

In addition to political reasons there are theoretical arguments which raise doubts about the potential of an international climate change agreement. The main obstacle for an international environmental agreement from a theoretical point of view is *stability*. Because of the public-good character of greenhouse gas abatement there is an inherent tendency for countries to free ride, ie it will generally pay-off not to join an agreement, since by doing so a country can enjoy the benefit of the treaty without having to participate in the costs. The free-rider problem and the possibility and size of stable, self-enforcing agreements have been the subject of several game theoretic papers (eg Barrett, 1993b, 1992c). A stable agreement is defined as a situation in which no additional country is willing to join an agreement, but at the same time there is no incentive for signatories to withdraw. The results of these models are often highly dependent on particular modelling assumptions, but on a whole they seem to imply that the size of a partial coalition will be rather small, even if it may possibly be widened through the use of side-payments (Carraro and Siniscalco, 1992).

The difficulty of achieving full international cooperation has raised the question of the usefulness of unilateral action taken by a subgroup of countries, particularly by the developed world. Unfortunately, it appears that, because of the limited share of OECD emissions on the total, unilateral action from industrialized countries alone will in the long run not be sufficient to significantly reduce warming rates. As Manne (1994, p 198) observes: 'From 2020 onwards there is no way to stabilise global emissions unless the developing nations are somehow induced to join an international agreement. Their growth is too important to be ignored'. This view is confirmed by Subak and Kelly (1993) who have estimated future warming rates under different cooperation scenarios. Their results are reproduced in Table 10.5. The table shows that a global commitment, even if confined to

Table 10.5 Global warming under different abatement commitments (Temperature Change by 2100, °C)

Scenario		Temperature change
Business as usual	IPCC (1992a) scenario IS92a	2.4°C
Current commitments	Annex 1: emissions at 1990 levels by 2000 and thereafter; other countries: business as usual	2.2°C
Toronto Conference	Annex 1: 20% cut in CO_2 emissions by 2005; other countries: business as usual	2.0°C
Global commitments	Emissions from all parties at 1990 levels by 2000 and thereafter	1.4°C
Stabilization scenario	Reductions of 60% CO_2 and 20% anthropogenic CH_4 and N_2O by 2100	1.0°C

Note: Annex 1 countries include OECD nations and economies in transition.

Source: Subak and Kelly (1993).

emission stabilization only, will in the long run be more effective than relatively stringent unilateral action by Annex 1 countries alone.

Carbon leakage

There is another reason why unilateral action may not be effective in the long run. It has to do with what is called *carbon leakage*. The Subak and Kelly calculations of Table 10.5 assume that unilateral action by one group of countries will not affect the emission decisions of others. In reality, this may not be the case. The direct effect of reduced emissions in one set of countries could be partly offset by increased emissions from non-signatories. This effect has been termed 'carbon leakage'.

Carbon leakage can be analysed in two different ways. The first uses a game theoretic approach (eg Hoel, 1991b). In this set-up leakage occurs as a consequence of free-riding. The non-abating players in the game profit from the unilateral action taken by their rivals in that the damage levels they are themselves exposed to are also reduced. The most adverse impacts thus being avoided, it may then become efficient for non-signatories to increase their own emissions. Hoel also shows that unilateral action prior to an agreement may actually be counter-productive and in the end lead to a less stringent protocol. This is because unilateral action will unfavourably move the fallback position of the abating player in the bargaining game. The situation without agreement now looks less threatening and a stringent protocol is pursued with less fervour. However, as Hoel himself points out, the result may be too pessimistic in that it neglects the possible psychological effects of a unilateral move onto thus far reluctant negotiators.

The second, and more prominent way to analyse unilateral action neglects damage considerations and concentrates on the general equilibrium effects of unilateral abatement.[8] In a multi-country general equilibrium set-up with international trade, leakage occurs through two channels. First there are price and income effects. The decrease in the demand for carbon-intensive products in the abating countries will lead to a fall in world market prices, which in turn will boost demand in non-abating countries. The price effect will partly be offset by an income effect as the earnings from exporting these goods decrease. Barrett (1992b) and Bohm (1993) have discussed ways to further offset the price effect, basically through a parallel reduction of both the supply and demand of affected goods. The second effect is a more direct trade effect. As a consequence of increased costs, carbon-intensive industries in constrained locations may suffer a loss in competitiveness and the production may shift from abating to non-abating regions.[9]

Carbon leakage is thus undesirable both from an environmental and

---

8 See Manne (1994), Pezzey (1992a), Oliveira-Martins et al (1992), Felder and Rutherford (1993), Rutherford (1993), Manne and Rutherford (1993) and Perroni and Rutherford (1991).

9 The impact of a carbon tax upon UK competitiveness has been analysed by Pezzey (1992b).

Table 10.6 Unilateral emissions abatement and carbon leakage (Leakage Rate in %)

Unilateral action by	2000	2010	2030	2050
Emission stabilization[a]				
United States only	2.4	0.5	−0.7	−0.2
Japan only	13.5	3.1	1.2	2.1
European Union only	11.2	5.5	2.9	2.2
Other OECD only	8.4	5.8	3.2	0.6
All OECD	2.4	−0.5	0.3	1.4
20% emission reduction[b]				
All OECD, trade in				
− oil only	8.1	0.6	−2.6	1.6
− oil, energy intensive products	13.0	14.8	21.8	17.7
− oil, gas, energy intensive products	18.2	20.5	28.1	30.8

Sources: a Oliveira-Martins et al (1992); b Manne (1994).

economic point of view. Unilateral action would not only be of little environmental benefit, it would also excessively punish the abating country on the international market. The size of the leakage effect in absolute terms is still unclear, however. Model predictions range from very modest to almost 100 per cent.[10] Everything else being equal, the leakage rate is higher, the larger the unilateral cut-back and the smaller the coalition of abating nations.

Unfortunately, the comparison of different results is rather difficult since the various studies all assume different policy scenarios. Results depend on several factors, most importantly on the amount of trade allowed in carbon intensive goods and fossil fuels. Manne (1994) points out that models using the Heckscher-Ohlin assumption of perfect substitutability between goods from different countries (eg Perroni and Rutherford, 1991; Rutherford, 1993; Pezzey, 1992a) obtain notably higher leakage rates than those using the 'Armington' hypothesis in which goods of different origin are imperfect substitutes (eg Oliveira-Martins et al, 1992). Other factors playing a role may include the supply elasticity of fossil fuels (Oliveira-Martins et al, 1992) and the modelling of the dynamics (Manne, 1994). To fully understand this important question and close the gap between the different model predictions a more careful and thorough model comparison would, however, be required. Table 10.6 reproduces results obtained with the OECD GREEN model (Oliveira-Martins et al, 1992) as well as calculations by Manne (1994).

---

10 The leakage rate is defined as the ratio of increased emissions by non-signatories relative to the reductions undertaken by abating nations. Thus, if the rest of the world increases CO_2 emissions by 3.5 mtC in response to a unilateral emissions cut of 35 mtC by the European Union the leakage rate is 10 per cent.

Chapter 11

Conclusions: Open Questions and State of the Art

Greenhouse economics is both a recent and quickly evolving field. Little of the material which currently constitutes the state of the art is more than three or four years old, and it is unlikely that contributions currently made will have a significantly longer lifetime. This book has surveyed the current level of the debate.

The discussion has reached a fairly sophisticated level with respect to the costs of abatement. Open questions mainly concern the gap between top down and bottom up models, and the discrimination between the different modelling specifications in general. Model comparison exercises are thus likely to continue. The discussion may also move in the direction of a more comprehensive approach, which will give more prominence to non-energy related abatement options and to the contribution of greenhouse gases other than CO_2. A wider view should also be adopted with respect to other existing market distortions and externalities. The double dividend and secondary benefit arguments clearly deserve more attention. They should however be treated in a more comprehensive context which considers all distortions equally and simultaneously. The current discussion suffers somewhat from a primacy of the greenhouse problem, in that improvements in other areas are seen as welcome side-effects of a carbon policy, but are not considered or sought in their own right.

The largest research input is probably required to complete the picture on global warming damage costs, an area which is still in its infancy. In many areas, but far from all, research will be able to found on a large scientific literature, which will contain at least a qualitative assessment. Further research is particularly required in three respects.

First, existing estimates have to be improved. The emphasis has to shift away from the agriculture and sea level rise sectors, where a comparatively high level of sophistication has already been achieved, to include other damage sectors, particularly non-market related damages like climate amenity, health and morbidity, and ecosystems loss, but also the impacts on, for example, the water industry. With respect to market-related damage, efforts will be needed to include indirect damage effects, ie to move from a partial equilibrium to an integrated approach.

Second, more research is needed to assess in more detail the damage in those regions which are most likely going to be the hardest hit: the developing countries. There are hardly any valuation studies for non-OECD countries, and such undertakings could therefore be of general methodological value.

Third, we have to overcome the concentration on the $2xCO_2$ benchmark. More research, both scientific and economic, is needed into the long-run effect of global warming and into the identification of potential thresholds. Economic damage modelling has to be raised onto the same level of sophistication as achieved in other sectors.

Improvements on the damage side will allow more reliable recommendations to be produced as to what the optimal policy response may be. Policy action will concern both greenhouse abatement and damage mitigation/adaptation. The economic community mostly seems to favour an approach based on cost-benefit considerations with respect to both concerns, and this for compelling reasons. This is not to say that a pure cost-benefit approach, as it is represented in the current generation of optimal abatement models, would be sufficient. It would not. The trade-off between costs and benefits also has to account for existing or remaining damage uncertainties, for irreversibilities, gradual learning, risk aversion and the possibility of climate surprises and catastrophes.

The question of how to hedge against climate uncertainties, and in particular of how to best deal with low probability/high damage events will have to be a main topic of future research in the area of optimal policy response. Another one will have to be the discounting question, given the high sensitivity of current results to this parameter. The question of intergenerational equity will also have to be tied with that of intragenerational equity, which dominates the international debate.

In terms of instruments, economists more or less in unison favour the use of market-based instruments, for the well-known efficiency reasons. The relatively new idea of joint implementation, which has gained policy relevance through its inclusion in the Climate Convention, will nevertheless deserve further analytical scrutiny as a more pragmatic alternative to tradeable emission permits. The policy debate will be more concerned about distributional questions, however, both domestically and with respect to international burden sharing.

Appendix 1

The Model of Chapter 4 in More Detail

This appendix describes the model underlying the marginal damage figures presented in Chapter 4. It is based on Fankhauser (1994a).

Greenhouse gases are so-called stock pollutants. The damage costs of a tonne of emission of gas i, S_i, are therefore a present value figure – the discounted sum of future incremental damage, $\partial D_t/\partial E_{i0}$,

$$S_i = \sum_{t=0}^{\tau} \frac{\partial D_t}{\partial E_{i0}} \cdot (1+\delta)^{-t} \tag{1}$$

where δ is the discount rate.

A model to estimate such a figure requires several elements. In particular, we need:

- Assumptions about future greenhouse gas emissions. Since the climate process is highly non-linear annual marginal damages will depend on future emission levels.
- A climate module transforming an increase in emissions in one period into incremental atmospheric concentration and then warming.
- A temperature-damage relationship.

In addition, we need to determine the appropriate discount rate. Because of the long time horizon, figures are particularly sensitive to this controversial parameter.

Future emissions

The model distinguishes between ten sources of emissions, including CO_2 from fossil fuel use and deforestation, four sources of methane emissions, N_2O and three CFCs. Business-as-usual (BAU) emissions are calibrated to mimic the IPCC (1992a) scenarios. In the case of CO_2 emissions from fossil fuel combustion, the emission growth rate is calculated as:

$$g_t = c_t + f_t + y_t + p_t \tag{2}$$

where c_t is the percentage change in carbon intensity (carbon emitted per energy unit), f_t the increase in energy efficiency (energy used per unit of GNP), y_t the change in per capita income and p_t the rate of population growth.

To take the possibility of future greenhouse gas abatement into account, BAU emissions are weighted against an abatement path. Emissions follow BAU with probability π, and with probability $(1-\pi)$ they are reduced through an abatement strategy. Emissions of gas i at time t, E_t^i, are therefore:

$$E_t^i = \pi \cdot BAU_t^i + (1 - \pi) \cdot STAB^i \tag{3}$$

where $STAB^i$ represents emissions under the abatement policy.

Accelerated abatement will not be costless, and weighted future income is therefore corrected for abatement costs, using a back-of-the-envelope method and cost figures suggested by Cline (1992a, 1994).

Atmospheric concentration

The accumulation of non-CO_2 emissions in the atmosphere is modelled through a set of equations of the form:

$$Q_t^i = \left(1 - \frac{1}{L_i}\right) \cdot Q_{t-1}^i + \beta_i E_t^i \tag{4}$$

That is, the atmospheric concentration of a gas, Q_t^i, is assumed to depreciate at a constant rate $1/L_i$, where L_i is the lifetime of gas i. β_i is a conversion parameter which transforms emissions (measured in tonnes) into concentrations (measured in parts per million by volume, ppm). In the case of methane the equation contains two decay factors, one representing atmospheric depreciation and the other the accumulation of CH_4 in the soil (Wigley and Raper, 1992).

The representation of the carbon cycle follows Maier-Reimer and Hasselmann (1987). In this formulation the carbon stock is modelled as a series of five boxes with constant, but different atmospheric lifetimes. That is, each box is modelled as in equation (4), and atmospheric concentration is a weighted sum of all five boxes (see also Lashof and Ahuja, 1990).

Radiative forcing

As a next step, increases in concentration have to be transformed into an increase in radiative forcing, that is into an increased flux of energy re-emitted from the atmosphere back to the earth. The relevant formulations for this step were taken from IPCC (1990a). For CO_2 the relationship between concentration and radiative forcing is logarithmic. In the case of CH_4 and N_2O, forcing depends on the square root of concentration levels. For CFCs the relationship is linear.

The forcing effect of CFCs is, however, diminished through an ozone feedback. Because CFCs react with stratospheric ozone, which is itself a greenhouse gas, they have an indirect cooling effect. The model neglects the potential cooling effect from sulphur emissions, on the other hand. Through an increased back-scattering of incoming solar radiation, sulphur aerosols may cause a reduction in observed warming levels (see IPCC, 1992a).

Temperature rise

The climate system is represented by a multi-stratum system consisting of three layers, the atmosphere, upper oceans and deep oceans. Increased radiative forcing warms up the atmosphere, which in turn heats the upper oceans, which then warm up the deep oceans. Both ocean layers impose a certain amount of thermal inertia on the system, which therefore only adjusts gradually to an increase in forcing. The system is represented by two partial adjustment equations (Schneider and Thompson, 1981; Nordhaus, 1992):

$$T_t^u = T_{t-1}^u + \frac{1}{R^u}\left[F_t - \lambda T_{t-1}^u - \frac{R^l}{\theta}(T_{t-1}^u - T_{t-1}^l)\right] \tag{5}$$

$$T_t^l = T_{t-1}^l + \frac{1}{R^l}\left[\frac{R^l}{\theta}(T_{t-1}^u - T_{t-1}^l)\right] \tag{6}$$

T_t^u and T_t^l are the temperature of the upper and lower layer of oceans, respectively (relative to the preindustrial level). R^u and R^l denote the thermal capacity of the layers, F_t the increase in radiative forcing at time t, and θ the transfer rate between upper and deep oceans. λ is a climate feedback parameter, related to the climate sensitivity of the system. That is, it indicates by how much temperature changes for a given increase in radiative forcing.

Annual damage

The representation of annual damage is calibrated around the $2xCO_2$ estimates of Chapter 3. Chapter 3 estimates the costs of global warming as would occur if a society with the economic structure of today was, at some point in the future, exposed to a temperature increase of $2.5°C$. This point estimate is expanded to a damage function of the form:

$$D_t = k_t \left(\frac{T_t^u}{\Lambda}\right)^\gamma \cdot (1 + \phi)^{t^* - \bar{t}} \tag{7}$$

where the two parameters and t^* represent two key assumptions of the $2xCO_2$ figures, *viz* the amount of warming associated with concentration doubling and the time at which it is assumed to occur. The values are set at $\Lambda = 2.5°C$ and $t^* = 2050$. Note that when $T_t^u = \Lambda$ and $\bar{t} = t^*$ annual damage in period t becomes $D_t = k_t$. k_t therefore represents the $2xCO_2$ estimates introduced in Chapter 3, adjusted for economic and population growth. The two parameters γ and ϕ determine damage outside the $2xCO_2$ benchmark.

Parameter γ defines the relationship between temperature and damage. If temperature rises by 1 per cent, damage rises by γ per cent. Little is known about the value of γ, although it can reasonably be assumed that $\gamma \geq 1$, ie that damage is convex in temperature. The work by Cline (1992a) and a poll of experts by Nordhaus (1993d, 1994) both suggest a value in the order of 1.3.

Simulation studies have typically used values between 1 and 3 (Nordhaus, 1991b, c, 1992, 1993a, b; Peck and Teisberg, 1992, 1993a, b).

A temperature rise of 3°C by 2100 is not the same as 3°C obtained in 2025. Damage may be considerably reduced if the system is given enough time to adapt. For example, under a sufficiently slow change, natural systems will be able to acclimatize or migrate to more favourable areas. In the case of economic damage, a slow rise in temperature may allow a gradual adjustment within the normal process of depreciation and reinvestment. While such considerations are not directly modelled, they are on an *ad hoc* basis embodied in parameter ϕ, a factor which augments impacts if they occur earlier than initially assumed (ie $2xCO_2$ is reached at time $\bar{t} < t^*$, and diminishes them if they are delayed.

This leaves the question of how damage changes as a consequence of economic development and population growth. Here we distinguish between market-based damage, affecting the national product, and non-market-based damage, such as people's valuation of a species or a warmer climate. The former is assumed to grow in proportion to GNP,

$$\frac{k_t^Y}{k_{t-1}^Y} = (1 + y_t^w + p_t) \tag{8}$$

where p_t denotes the rate of growth of population, as introduced earlier. y_t^w is the rate of growth of per capita income, taking into account the costs of possible carbon abatement, as well as potential warming impacts on growth.

For non-market impacts the formula is slightly more complex. If people's willingness to pay for non-market goods was constant over time, non-market damage would grow at the same rate as population. However, this will not be the case. As people become richer over time, their willingness to pay will also change. Defining ε_Y as the income elasticity of people's willingness to pay to avoid non-market damage, the growth factor of non-market damage should therefore be written as:

$$\frac{k_t^p}{k_{t-1}^p} = (1 + \epsilon_Y \cdot y_t^w + p_t) \tag{9}$$

Estimates of ε_Y for environmental goods in general are surveyed by Pearce (1980). Without finding any conclusive results, the survey seems to suggest an elasticity value in the order of $\varepsilon_Y = 1$. The initial values for k_0^Y and k_0^P were taken from Chapter 3, from where we derived that 62 per cent of the damage is non-market related and 38 per cent market related.

Discounting

With respect to discounting, the model follows the standard consumption equivalent technique developed by, among others, Arrow (1966), Arrow and Kurz (1970), Bradford (1975), Marglin (1963a and b), and Dasgupta et al (1972).[1] In the context of global warming this method has been used by, for example, Cline (1992a).

In an ideal world without distortions, damage flows could either be discounted at the social rate of time preference (the consumer rate of discount), or at the marginal product of capital (the opportunity cost of investment). In a first best world the two rates would be equal. In reality they will in general be different. The appropriate rate of discount then depends on the source of the funds, or, in the case of greenhouse damage, on whether climate change affects investment or consumption goods. Consumption-based damages should be discounted at the social rate of time preference, and investment-related damage at the rate of return on capital.

Alternatively, all flows could first be transformed into consumption units and then discounted uniformly at the social rate of time preference. This is the method followed here. In modelling this process two parameters, therefore, have to be defined: the social rate of time preference, δ; and the shadow value of capital, v, which transforms investment into consumption units.

Under the usual assumption of a CRRA (constant relative rate of risk aversion) utility function the social rate of time preference, δ, can be written as (see Lind, 1982)

$$\delta_t = \rho + \omega y_t^w \qquad (10)$$

The social rate of time preference thus consists of two elements. The first one, denoted ρ, is called the pure rate of time preference. It deals with the impatience of consumers and reflects their preference of immediate over postponed consumption. The second element has to do with decreasing marginal utility of income. A decreasing marginal utility of income implies that a dollar of additional income will create less additional utility the higher the initial income level. Consequently, if income is rising over time, future impacts should be valued less. The income elasticity of marginal utility is denoted by ω. It has to be multiplied by the rate of growth of per capita income, y_t^w. Note that since income growth may vary over time the rate of discount changes over time as well. Growth may again be affected by future abatement policies or by greenhouse impacts.

The most controversial element in this equation is the pure rate of time preference, see the discussion in Chapter 8. For the main results in Chapter 4, ρ is a random variable with upper and lower bounds of 0 and 0.03, respectively, and a best guess of 0.005.

The second element which needs to be defined is the shadow value of capital, v. The shadow price of capital basically represents the present value of the future consumption stream associated with a $1 investment, discounted at the social rate of time preference (Lind, 1982). Here we use a simplified method suggested by Cline (1992a).

1 See eg Lind (1982), Hanley (1992) and Pearce and Nash (1981), for surveys.

Suppose a \$1 investment project yields an annual payoff of A over a lifetime of N years. The present value consumption stream is then (omitting the time subscript on δ):

$$v = \sum_{\tau=1}^{N} A(1+\delta)^{-\tau} \qquad (11)$$

If the project has an internal rate of return equal to the rate of return on capital, r, it will further satisfy the equation,

$$-1 + \sum_{\tau=1}^{N} A(1+r)^{-\tau} = 0 \qquad (12)$$

Solving for A, and substituting back yields, after some further rearranging:

$$v_t = \frac{r}{1 - (1+r)^{-N}} \cdot \frac{1 - (1+\delta_t)^{-N}}{\delta_t} \qquad (13)$$

Whether or not an expenditure has to be transformed depends on its character, as outlined above. Climate impacts affecting investment have to be transformed, while reductions in consumption need not, as they are already expressed in the correct units. Based on the currently observed average savings rate, we assume that 20 per cent of market-based damage is investment related, ie has to be multiplied by a factor v. 80 per cent of income-related damage and all non-market damage is assumed to be consumption based. The $2\text{x}CO_2$ damage variable k_t can then be rewritten as:

$$k_t = (k_t^P + 0.8k_t^Y) + 0.2v_t k_t^Y \qquad (14)$$

Appendix 2

The Model of Chapter 5 in More Detail

Appendix 2 summarizes the model of Fankhauser (1994b), which was used for the calculations of Chapter 5. To recapitulate: we seek to minimize the present value costs of SLR over time. There are three cost elements: protection costs (PC), dryland loss (DL) and wetland loss (WL). Costs are minimized with respect to the percentage of coasts protected, L. A second decision variable is the additional height added to protection measures in each period, denoted h_t. The problem is thus a combination of a static optimization (with respect to variable L), and a dynamic optimization (in the case of h_t). It can be expressed as follows:

$$\min_{L, h_t} Z = \int_0^{\tau} [PC(L, h_t, G_\tau) + DL_t(L, S_t) + WL_t(L, S_t)] e^{-\delta t} dt \tag{1}$$

subject to

$$
\begin{aligned}
&\dot{G}_t = h_t \\
&G_t \geq S_t \\
&h_t \geq 0 \\
&0 \leq L \leq 1 \\
&G_0 \text{ given, } G_\tau \text{ free}
\end{aligned}
\tag{2}
$$

S_t is the exogenously given SLR path and G_t the height of protection measures at time t. δ is the discount rate. The requirement $G_t \geq S_t$ reflects the fact that for the sea wall to be effective it has to be at least as high as the sea level.[1]

The discount rate is determined in same way as in the model of Chapter 4 (see Appendix 1). The variables PC, DL_τ and WL_τ are defined as follows.

Protection costs

Based on other estimates (IPCC, 1990c; Titus et al, 1991) we assume that protection costs rise exponentially with height, but are linear in the length of

---

1 In a model allowing for storm surges, ie where S_t fluctuates over time, the optimal height would be determined in a trade-off between the costs of construction and the expected benefits of flooding damage avoided, calculated as the damage per flood event times the probability of occurrence. In the present formulation S_t is known with certainty, whence we can simply impose $G_t \geq S_t$.

protection measures. Costs are exponential in height basically because a higher sea wall will, for stability reasons, also have to be more massive.

Since the sea level is rising only gradually over time, construction can be spread over several periods. What we are therefore interested in is a stream of annual costs, PC. We assume that, once the final height of the wall is determined, it makes no difference whether an additional layer of bricks is added today or tomorrow, as long as we abstract from discounting effects and neglect fixed costs (eg hiring and firing costs): a brick is a brick is a brick. While simplistic, the 'a brick is a brick' assumption is central to the model. It assures that PC is linear in h_t, which will later allow us to seek a simple, sequential solution to the model.

Annual protection costs are then a function of the length of the wall, its ultimate height, G_τ, and the height added in each period, h_t

$$PC(L, h_t, G_\tau) = LK \cdot \varphi \cdot G_\tau^{\gamma-1} \cdot h_t \tag{3}$$

where K denotes the length of threatened coastline, and $L \cdot K$ therefore the length of coastlines protected.

Dryland loss

SLR damage from dryland loss depends on two aspects: on the amount of land lost and on its value. For the former we assume that a land area of ψ will be lost per cm of SLR and km of coastline.[2] Given that a fraction L of all coastlines will be protected, a SLR of S_t will inundate an area of $(1-L)K\psi S_t$. To get the monetary damage, this value is multiplied by the annual return on land, R_t. R_t is the opportunity cost of lost land, ie it denotes the return an area would have yielded, had it not been inundated. The dryland loss at time t is therefore:

$$DL_t(L, S_t) = (1 - L)K \cdot \Psi S_t \cdot R_t(L) \tag{4}$$

The return on land, R_t, is simply the value of lost land multiplied by the rate of return on capital r. Since more valuable areas are protected with priority, the average return on unprotected land, R_t will decrease with increasing protection efforts, as explained in Chapter 5. To keep the linear structure of the model, the relationship between R_t and L is specified as:

$$R_t(L) = r \cdot x_t(1 - L) \tag{5}$$

For $L = 0$, ie if no protection takes place, the average value of unprotected coasts equals the overall expected value for land, x_t, and the annual return is consequently rx_t. As L increases, the value of unprotected areas decreases and approaches zero for the last unprotected coastal strip. The value of land,

---

2 Strictly, this linear relationship only applies to flat coasts. In the special case of soft rock cliffs, which are threatened from increased erosion, the underlying processes are more complex. In most cases, though, soft cliffs are preceded by beaches (created through previous erosion) for which the linearity assumption holds.

x_t, is assumed to increase over time in proportion to economic growth as land becomes more scarce relative to income.[3]

Wetland loss

As outlined in Chapter 5, wetlands are in principle able to adjust to SLR by migrating inland, but may be too slow to adjust fully. In addition, backward migration is only possible if there are no obstacles in the way, and may therefore be limited to unprotected areas.

Suppose that wetlands are able to migrate at a speed of α. At time t coastal wetlands will then have migrated by αt and will have grown by an area of $W(1-L)\alpha t$, where W denotes the length of coastal wetlands.[4] At the same time some land will have been lost through inundation. We have[5]

$$WL_t(L, S_t) = [\Psi S_t - (1 - L)\alpha t]W \cdot R_t^W \tag{6}$$

As opposed to the dryland case, the return on wetlands, R_t^W, does not depend on the degree of protection. Recall that equation (5) was driven by the assumption that high value dryland will be protected first. No similar assumption holds for wetlands, which are affected only indirectly. Assuming that there is no correlation between dryland and wetland values, the expected value of affected wetlands will therefore be equal to the mean value of coastal wetlands. Again we assume that land values rise in proportion to economic growth.

The optimal height of protection measures

From the fact that the optimization problem (1) is linear in h_t it follows that, while the optimal value for L, L^*, depends on h_t, h_t^* will not depend on L. The model can thus be solved sequentially in a two-step process. First we derive the optimal strategy with respect to h_t, h_t^*. In a second step we then calculate L^*, given that $h_t = h_t^*$.

Under the assumption that 'a brick is a brick', the optimization of (1) with respect to h_t is fairly straightforward. h_t and G only occur in the expression on protection costs, PC. The question thus is how to raise the sea wall at minimum costs without violating the constraints. In principle, the lower the height of the wall, the better. However, in the terminal period we require $G_\tau \geq S_\tau$. Since less height is better, the constraint will be strictly binding, $G_\tau = S_\tau$. The ultimate height of the wall is thus set by the amount of SLR expected

---

3 Arguably, the rise in value of land in general could partly be offset by a decrease in the valuation of coastal areas, as they become increasingly vulnerable over time.
4 Strictly, W denotes the length of *marsh wetlands*. Freshwater wetlands cannot migrate and will be lost. We can think of this loss as being included in dryland loss.
5 Note that the rate of inland migration cannot exceed the rate of inundation (wetlands do not expand into dryland areas). Neither can the wetland loss exceed the total amount of wetlands. Equation (6) thus only holds for $0 \leq (\Psi S_t - \alpha t) \leq 1$. The condition holds for all numerical specifications considered in Chapter 5.

in the terminal period. This leaves the question of the optimal timing of construction work. In the absence of fixed costs and under the assumption that 'a brick is a brick', construction tomorrow is a perfect substitute of construction today, ie there are no gains connected with a one-off erection of measures. The discounting effect will therefore see to it that construction is carried out as late as possible, constrained only by the requirement $G_t \geq S_t$.

The optimal strategy with respect to the height of defence measures will therefore be not to raise the installations as long as they are higher than the current sea level or if the sea level is falling, and to increase them at the same pace as the sea is rising otherwise.

The optimal length of coastal protection

With respect to variable L, the proportion of coastlines to be protected, the dynamic character of the optimization problem is of no relevance. In fact, once h_t^* is determined the model becomes completely static. To simplify the notation we introduce the following four variables.

Define PC^{pv} as the present value protection costs under full protection, ie for $L = 1$.

$$PC^{pv}(h_t, G_\tau) \equiv \int_0^\tau PC(1, h_t, G_\tau)e^{-\delta t} dt \qquad (7)$$

Define DL^{pv} as the present value of dryland loss damage under no protection, i.e. for $L = 0$,

$$DL^{pv}(S_t) \equiv \int_0^\tau DL_t(0, S_t)e^{-\delta t} dt \qquad (8)$$

Finally we define WG^{pv} as the present value gain from the inland migration of wetlands under $L = 0$ (ie if full adaptation is possible), and WL^{pv} as the present value damage from wetland loss. This implies:

$$WL^{pv}(S_t) - WG^{pv} = \int_0^\tau WL(0, S_t)e^{-\delta t} dt \qquad (9)$$

Substituting in, expression (1) becomes:

$$\min_{L,h_t} Z = [L \cdot PC^{pv} + (1 - L)^2 DL^{pv} + WL^{pv} - (1 - L)WG^{pv}] \qquad (10)$$

subject to (2). Taking advantage of the fact that h_t^* is independent of L, and setting $h_t = h_t^*$, we derive

$$L^{opt} = 1 - \frac{1}{2}\left(\frac{PC^{pv} + WG^{pv}}{DL^{pv}}\right) \qquad (11)$$

Finally, we have to remember that L is bounded from below and above. The upper bound condition, $L^{opt} \leq 1$, presents no problem. It will never be violated as long as the three present value terms are positive. To secure the lower bound condition, $L^{opt} \geq 0$, we rewrite equation (11) as:

$$L^* = \begin{cases} 0 & \text{if } L^{opt} < 0 \\ L^{opt} & \text{otherwise} \end{cases} \qquad (12)$$

This is the expression used in Chapter 5.

Bibliography

Adams, R M, C Rosenzweig, R M Peart, J T Ritchie, B A McCarl, J D Glyer, R B Curry, J W Jones, K J Boote, and L H Allen Jr (1990), Global Climate Change and US Agriculture, in *Nature* 345(17 May): 219–23.

Adger, N and S Fankhauser (1993), Economic Analysis of the Greenhouse Effect: Optimal Abatement Level and Strategies for Mitigation, in *International Journal of Environment and Pollution* 3(1–3): 104–19.

Agarwal, A and S Narain (1991), *Global Warming in an Unequal World. A Case of Environmental Colonialism*, Mimeo, Centre for Science and Environment, New Delhi.

Alcamo, J, G J J Kreileman, M Krol, and G Zuidema (1993), *Modelling The Global Society-Biosphere-Climate System: Part 1: Model Description and Testing*, Mimeo, National Institute of Public Health and Environmental Protection (RIVM), Bilthoven, The Netherlands.

Alfsen, K, H Birkelund, and M Aaserud (1993), *Secondary Benefits of the EC Carbon/ Energy Tax*, Research Department Discussion Paper No 104, Statistics Norway, Oslo.

Alfsen, K H, A Brendemoen, and S Glomsrød (1992), *Benefits of Climate Policies: Some Tentative Calculations*, Discussion Paper No 69, Norwegian Central Bureau of Statistics, Oslo.

Ali, S I and S Haq (1990), *International Sea Level Rise: National Assessment of Effects and Possible Responses for Bangladesh*, Unpublished Report to the University of Maryland, College Park.

Amano, A (1994), *Estimating Secondary Benefits of Limiting CO_2 Emissions in the Asian Region*, Mimeo, School of Business Administration, Kobe University.

Anderson, D and R Williams (1993), *The Cost Effectiveness of GEF Projects*, Working Paper No 6, Global Environment Facility, Washington DC.

Arrhenius, S A (1896), On the influence of carbonic acid in the air upon temperature on the ground, in *Philosophical Magazine* 41(251): 237–276.

Arrhenius, S A (1903), *Lehrbuch der kosmischen Physik*, 2 volumes, Leipzig: Hirzel.

Arrhenius, S A (1908), *Worlds in the Making*, translated by H Borns, New York: Harper.

Arrow, K J (1966), Discounting and Public Investment Criteria, in A V Kneese and S C Smith (eds), *Water Research*, Baltimore: Johns Hopkins University Press.

Arrow, K J and A C Fisher (1974), Environmental Preservation, Uncertainty and Irreversibility, in *Quarterly Journal of Economics* 88: 312–19.

Arrow, K J and M Kurz (1970), *Public Investment, the Rate of Return and Optimal Fiscal Policy*, Baltimore: Johns Hopkins University Press.

Ayres, R and J Walter (1991), The Greenhouse Effect: Damages, Costs and Abatement, in *Environmental and Resource Economics* 1: 237–70.

Ballard, C L, J B Shoven, and J Whalley (1985), General Equilibrium Computations of the Marginal Welfare Costs of Taxes in the United States, in *American Economic Review* 75(1): 128–38.

Barker, T (1993), *Secondary Benefits of Greenhouse Gas Abatement: The Effects of a UK Carbon/Energy Tax on Air Pollution*, Energy Environment Economy Modelling Discussion Paper No 4, Department of Applied Economics, University of Cambridge.

Barker, T, S Baylis, and P Madsen (1993), A UK Carbon/Energy Tax: The Macroeconomic Effects, in *Energy Policy* 21(3): 296–308.

Barrett, S (1992a), 'Acceptable' Allocations of Tradeable Carbon Emission Entitlements in a Global Warming Treaty, in UNCTAD, *Combating Global Warming. Study on a Global System of Tradeable Carbon Emission Entitlements*, Geneva: UNCTAD.

Barrett, S (1992b), *Convention on Climate Change. Economic Aspects of Negotiations*, Paris: OECD.

Barrett, S (1992c), *Self Enforcing International Environmental Agreements*, Global Environmental Change Working Paper GEC 92–34, Centre for Social and Economic Research on the Global Environment, University of East Anglia, Norwich, and University College London.

Barrett, S (1993a), *A Strategic Analysis of 'Joint Implementation' Mechanisms in the Framework Convention on Climate Change*, Paper prepared for UNCTAD, Draft, London Business School and CSERGE, London.

Barrett, S (1993b), *Heterogeneous International Environmental Agreements*, Global Environmental Change Working Paper GEC 93–20, Centre for Social and Economic Research on the Global Environment, University of East Anglia, Norwich, and University College London.

Berger, K, O Fimreite, R Golombek, and M Hoel (1992), The Oil Market and International Agreements on CO_2 Emissions, in *Resources and Energy* 14(4): 315–36.

Bergman, L (1991), General Equilibrium Effects of Environmental Policy: A CGE Modelling Approach, in *Environmental and Resource Economics* 1(1): 43–62.

Berz, G (1990), Natural Disasters and Insurance/Reinsurance, in *UNDRO News*, January/February: 18–19.

Bigford, T E (1991), Sea Level Rise, Nearshore Fisheries, and the Fishing Industry, in *Coastal Zone Management* 19: 417–37.

Binkley, CS (1988), A Case Study of the Effects of CO_2-Induced Climatic Warming on Forest Growth and the Forest Sector: B Economic Effects on the World's Forest Sector, in M L Parry, T R Carter, and N T Konijn (eds), *The Impacts of Climate Variations on Agriculture. Volume 1: Assessment in Cool, Temperate and Cold Regions*, Dordrecht: Kluwer.

Bishop, R C (1978), Endangered Species and Uncertainty: The Economics of a Safe Minimum Standard, in *American Journal of Agricultural Economics* 60(1): 10–18.

Birdsall, N and A Steer (1993), Act Now on Global Warming – But Don't Cook the Books, in *Finance & Development* 30(1): 6–8.

Blitzer, C R, R S Eckaus, S Lahiri, and A Meeraus (1992), *Growth and Welfare Losses from Carbon Emissions Restrictions: A General Equilibrium Analysis for Egypt*, Background Paper for the World Development Report, The World Bank, Washington, DC.

Boero, G, R Clarke, and L A Winters (1991), *The Macroeconomic Consequences of Controlling Greenhouse Gases: A Survey*, Environmental Economics Research Series, UK Department of the Environment, London: HMSO.

Bohm, P (1993), Incomplete International Cooperation to Reduce CO_2 Emissions: Alternative Policies, in *Journal of Environmental Economics and Management* 24(3): 258–71.

Bohm, P (1994), *On the Feasibility of Joint Implementation of Carbon Emissions Reductions*, Department of Economics Discussion Paper No 94–05, University of Birmingham.

Bradford, D F (1975), Constraints on Government Investment Opportunities and the Choice of Discount Rate, in *American Economic Review* 65(50): 887–99.

Brechling, V and S Smith (1994), Household Energy Efficiency in the UK, in *Fiscal Studies* 15(2): 44–56.

Brinner, R E, M G Shelby, J M Yanchar, and A Cristofaro (1991), Optimizing Tax Strategies to Reduce Greenhouse Gases without Curtailing Growth, in *Energy Journal* 12(4): 1–14.

Broome, J (1992), *Counting the Costs of Global Warming*, Cambridge: White Horse Press.

Bryant, E A (1991), *Natural Hazards*, Cambridge: Cambridge University Press.

Burniaux, J M, J P Martin, and J Oliveira-Martins (1992a), The Effect of Existing Distortions in Energy Markets on the Costs of Policies to Reduce CO_2 Emissions: Evidence from GREEN, in *OECD Economic Studies* No 19/Winter.

Burniaux, J M, G Nicoletti, and J Oliveira-Martins (1992b), GREEN. A Global Model for Quantifying the Costs of Policies to Curb CO_2 Emissions, in *OECD Economic Studies* No 19/Winter.

Canadian Climate Program Board (1988a), *Implications of Climate Change for Downhill Skiing in Quebec*, Climate Change Digest Report, CCD 88–03, Downsview, Ontario.

Canadian Climate Program Board (1988b), *Implications of Climate Change for Tourism and Recreation in Ontario*, Climate Change Digest Report, CCD 88–05, Downsview, Ontario.

Carraro, C and D Siniscalco (1992), *Strategies for the International Protection of the Environment*, Nota di Lavoro, 4.92, Fondazione ENI Enrico Mattei, Milan.

Chandler, W U (1990), *Carbon Emission Control Strategies: Case Studies in International Cooperation*, Baltimore: World Wildlife Fund and Conservation Foundation.

Chichilnisky, G, G Heal, and D Starrett (1993), *International Emission Permits: Equity and Efficiency*, Mimeo, Columbia University and Columbia Business School, New York.

Clarke, R (1993), Energy Taxes and Subsidies: Their Implications for CO_2 Emissions and Abatement Costs, in *International Journal of Environment and Pollution* 3(1–3): 168–178.

Climate Research Unit (1992), *A Scientific Description of the ESCAPE Model*, Mimeo, Climate Research Unit, University of East Anglia, Norwich.

Cline, W (1991), Scientific Basis of the Greenhouse Effect, in *Economic Journal* 101(407): 904–919.

Cline, W R (1992a), *The Economics of Global Warming*, Washington, DC: Institute for International Economics.

Cline, W R (1992b), *Optimal Carbon Emissions over Time: Experiments with the Nordhaus DICE Model*, Mimeo, Institute for International Economics, Washington, DC.

Cline, W R (1993a), *Modelling Economically Efficient Abatement of Greenhouse Gases*, Paper presented at the United Nations University Conference on 'Global Environment, Energy and Economic Development', September 1993, Tokyo.

Cline, W R (1993b), Give Greenhouse Abatement a Fair Chance, in *Finance & Development* 30(1): 3–5.

Cline, W R (1993c), *The Impacts of Global Warming on the United States: A Survey of Recent Literature*, Mimeo, Institute for International Economics, Washington, DC.

Cline, W R (1994), The Costs and Benefits of Greenhouse Abatement: A Guide to Policy Analysis, in OECD/IEA, *The Economics of Climate Change*, Paris: OECD.

COHERENCE (1991), *Cost Effectiveness Analysis of CO_2 Reduction Options*, Synthesis Report, Commission of the European Community, DG XII, Brussels.

Daily, G C, P R Ehrlich, H A Mooney, and A H Ehrlich (1991), Greenhouse Economics: Learn before you Leap, in *Ecological Economics* 4: 1–10.

Dasgupta, P, S A Marglin and A K Sen (1972), *Guidelines for Project Evaluations*, New York: UNIDO.

den Elzen, M G J and J Rotmans (1992), The Socio-Economic Impact of Sea-Level Rise on the Netherlands: A Study of Possible Scenarios, in *Climatic Change* 20: 169–195.

Dowlatabadi, H and M G Morgan (1993), A Model Framework for Integrated Studies of the Climate Problem, in *Energy Policy* 21(3): 209–21.

Dudek, D J and A LeBlanc (1990), Offsetting new CO_2 Emissions: A Rational First Greenhouse Policy Step, in *Contemporary Policy Issues* VIII(July): 29–42.

Easterling III, W E, P R Crosson, N J Rosenberg, M S McKenney, L A Katz and K M Lemon (1993), Agricultural Impacts of and Responses to Climate Change in the Missouri-Iowa-Nebraska-Kansas (MINK) Region, in *Climatic Change* 24(1/2): 23–62.

Economist, The (1991), *Water Prices*, Business this Week, Issue of 17 August 1991.

Edmonds, J A and D W Barns (1991), *Factors Affecting the Long-Term Cost of Global Fossil Fuel CO_2 Emissions Reductions*, Mimeo, Pacific Northwest Laboratory, Washington, DC.

Edmonds, J A, H M Pitcher, D Barns, R Baron, and M A Wise (1992), *Modelling Future Greenhouse Gas Emissions: The Second Generation Model Description*, Mimeo, Pacific Northwest Laboratory, Washington, DC.

Edmonds, J A, H M Pitcher, N J Rosenberg, and T M L Wigley (1993), *Design for the Global Change Assessment Model GCAM*, Paper presented at the IIASA Workshop on 'Integrative Assessment of Mitigation, Impacts and Adaptation to Climate Change', October 1993, Laxenburg, Austria.

Edmonds, J A and J Reilly (1983), Global Energy and CO_2 to the Year 2050, in *Energy Journal* 4(3): 21–47.

Ekins, P (1994), *Rethinking the Costs of Global Warming*, Department of Economics Discussion Paper No 94–07, University of Birmingham.

Emanuel, K A (1987), The Dependence of Hurricane Intensity on Climate, in *Nature* 326: 483–5.

Eyckmans, J, S Proost, and E Schokkaert (1993), Efficiency and Distribution in Greenhouse Negotiations, in *Kyklos* 46(3): 363–98.

Fankhauser, S (1994a), The Social Costs of Greenhouse Gas Emissions: An Expected Value Approach, in *Energy Journal* 15(2): 157–84.

Fankhauser, S (1994b), Protection vs Retreat. The Economic Costs of Sea Level Rise, in *Environment and Planning A*, forthcoming.

Fankhauser, S and S Kverndokk (1992), *The Global Warming Game – Simulations of a CO_2 Reduction Agreement*, Global Environmental Change Working Paper GEC 92–10, Centre for Social and Economic Research on the Global Environment, University College London and University of East Anglia, Norwich.

Felder, S and T Rutherford (1993), Unilateral CO_2 Reductions and Carbon Leakage: The Consequences of International Trade in Oil and Basic Materials, in *Journal of Environmental Economics and Management* 25(2): 162–76.

Fisher, A C and W M Hanemann (1993), Assessing Climage Change Risks: Valuation of Effects, in J Darmstaedter and M Toman (eds), *Assessing Surprises and Non-Linearities in Greenhouse Warming*, Washington, DC: Resources for the Future.

Food and Agricultural Organisation (1991a), *Yearbook Production*, Vol 44, Rome: FAO.

Food and Agricultural Organisation (1991b), *Yearbook of Fishery Statistics and Landings*, Vol 68, Rome: FAO.

Gaskins, D W and J P Weyant (1993a), Modelling Comparisons of the Costs of Reducing CO_2 Emissions, in *American Economic Review, Papers and Proceedings* 83(2): 318–23.

Gaskins, D W and J P Weyant (1993b), *Reducing Carbon Emissions from the Energy Sector: Cost and Policy Options*, Cambridge, MA: MIT Press.

Gleick, P H and E P Maurer (1990), *Assessing the Costs of Adapting to Sea Level Rise. A Case Study of San Francisco Bay*, Mimeo, Pacific Institute for Studies in Development, Environment and Security, and Stockholm Environment Institute, Berkeley, CA and Stockholm.

Glomsrød, S, H Vennemo, and T Johnsen (1992), Stabilisation of Emissions of CO_2: A Computable General Equilibrium Assessment, in *Scandinavian Journal of Economics* 94(1): 53–69.

Goulder, L H (1993), *Effects of Carbon Taxes in an Economy with Prior Tax Distortions: An Intertemporal General Equilibrium Analysis*, Mimeo, Stanford University and National Bureau of Economic Research, Stanford, CA.

Grubb, M (1989), *The Greenhouse Effect: Negotiating Targets*, London: Royal Institute of International Affairs.

Grubb, M (1990), *Energy Policies and the Greenhouse Effect. Volume 1: Policy Appraisal*, London: Royal Institute of International Affairs.

Grubb, M. (1993), The Costs of Climate Change: Critical Elements, in Y. Kaya, N. Nakićenović, W D Nordhaus and F. Toth (eds), *Costs, Impacts and Benefits of CO_2 Mitigation*, IIASA Collaborative Paper Series, CP93-2, Laxenburg, Austria.

Grubb, M, J Edmonds, P ten Brink, and M Morrison (1994), The Costs of Limiting Fossil Fuel CO_2 Emissions: A Survey and Analysis, in *Annual Review of Energy and Environment* 18: 397–478.

Grubb, M, M Koch, A Munson, F Sullivan, and K Thomson (1993), *The Earth Summit Agreements. A Guide and Assessment*, London: Royal Institute of International Affairs.

Grubb, M, J Sebenius, A Magalhaes and S Subak (1992), Sharing the Burden, in I M Mintzer (ed), *Confronting Climate Change, Risks, Implications and Responses*, Cambridge: Cambridge University Press.

Gyourko, J and J Tracy (1991), The Structure of Local Public Finance and the Quality of Life, in *Journal of Political Economy* 99(4): 774–806.

Haines, A and C Fuchs (1991), Potential Impacts on Health of Atmospheric Change, in *Journal of Public Health Medicine* 13(2): 69–80.

Hall, S, N. Mabey, and C Smith (1994), *Macroeconomic Modelling of International Carbon Tax Regimes*, Department of Economics Working Paper No 94–08, University of Birmingham.

Handel, M D and J S Risbey (1992), *An Annotated Bibliography on Greenhouse Effect Change*, Report No 1 (Reprinted with Updates), Center for Global Change Science, MIT, Cambridge, MA.

Hanley, N (1992), Are there Environmental Limits to Cost Benefit Analysis? in *Environmental and Resource Economics* 2(1): 33–60.

Harrod, R (1948), *Towards a Dynamic Economy*, London: Macmillan.

Heal, G (1984), Interactions between Economy and Climate: A Framework for Policy Design under Uncertainty, in *Advances in Applied Microeconomics* 3: 151–68.

Hervik, A, M Risnes and J Strand (1987), Implicit Costs and Willingness-to-pay for Development of Water Resources, in A J Carlsen (ed), *Proceedings – UNESCO Symposium on Decision Making in Water Resources Planning*, 5–7 May 1986, Oslo.

Hoch, I and J Drake (1974), Wages, Climate and the Quality of Life, in *Journal of Environmental Economics and Management* 1: 265 95.

Hoel, M (1991a), Efficient International Agreements for Reducing Emissions of CO_2, in *Energy Journal* 12(2): 93–107.

Hoel, M (1991b), Global Environment Problems: The Effects of Unilateral Action Taken by One Country, in *Journal of Environmental Economics and Management* 20(1): 55–70.

Hoel, M (1992), Carbon Taxes: An International Tax or Harmonised Domestic Taxes? in *European Economic Review* 36: 400–406.

Hoel, M and I Isaksen (1993), *Efficient Abatement of Different Greenhouse Gases*, Memorandum No. 5, Department of Economics, University of Oslo.

Hoeller, P and J Coppel (1992), Energy Taxation and Price Distortions in Fossil Fuel Markets: Some Implications for Climate Change Policy, in OECD, *Climate Change. Designing a Practical Tax System*, Paris: OECD.

Hoeller, P, A Dean, and M Hayafuji (1992), *New Issues, New Results: The OECD's Second Survey of The Macroeconomic Costs of Reducing CO_2 Emissions*, Economic Department Working Paper No 123, OECD, Paris.

Hoeller, P, A Dean, and J Nicolaisen (1991), Macroeconomic Implications of Reducing Greenhouse Gas Emissions: A Survey of Empirical Studies, in *OECD Economic Studies* No 16/Spring.

Hogan, W W and D W Jorgenson (1991), Productivity Trends and the Cost of Reducing CO_2 Emissions, in *Energy Journal* 12(1): 67–85.

Hohmeyer, O (1988), *Social Costs of Energy Consumption: External Effects of Electricity Generation in the Federal Republic of Germany*, Berlin: Springer.

Hope, C, J Anderson, and P Wenman (1993), Policy Analysis of the Greenhouse Effect. An Application of the PAGE Model, in *Energy Policy* 21(3): pp 327–38.

Hourcade, J C (1993), Modelling Long Run Scenarios: Methodological Lessons from a Prospective Study on a Low CO_2 Intensive Country, in *Energy Policy* 21(3): 309–26.

Howarth, R B and P A Monahan (1992), *Economics, Ethics, and Climate Policy*, Mimeo, Lawrence Berkeley Laboratory, Berkeley, CA.

Howarth, R B and R B Norgaard (1993), Intergenerational Transfers and the Social Discount Rate, in *Environmental and Resource Economics* 3(4): 337–58.

Ingham, A, A Ulph, and D Ulph (1993), *Carbon Taxes and Energy Markets*, Nota di Lavoro, 2.93, Fondazione ENI Enrico Mattei, Milan.

Intergovernmental Panel on Climate Change (1990a), *Climate Change, the IPCC Scientific Assessment.* Report from Working Group I, Cambridge: Cambridge University Press.

Intergovernmental Panel on Climate Change (1990b), *Climate Change, the IPCC Impacts Assessment.* Report from Working Group II, Canberra: Australian Govenment Publishing Service.

Intergovernmental Panel on Climate Change (1990c), *Strategies for Adaptation to Sea Level Rise.* Report of the Coastal Zone Management Subgroup, The Hague.

Intergovernmental Panel on Climate Change (1992a), *Climate Change 1992. The Supplementary Report to the IPCC Scientific Assessment*, Cambridge: Cambridge University Press.

Intergovernmental Panel on Climate Change (1992b), *Global Climate Change and the Rising Challenge of the Sea. Supporting Document for the IPCC Update 1992*, Geneva: WMO and UNEP.

International Energy Agency (1991), *Energy Statistics and Balances of Non-OECD Countries*, 1988–1989, Paris: OECD.

International Energy Agency (1992), *Energy Prices and Taxes.* Third Quarter 1991, Paris: OECD.

International Energy Agency (1994), *Climate Change Policy Initiatives 1994 update*, Vol I, Paris: OECD.

Jäger, J (1992), From Conference to Conference, in *Climatic Change* 20: iii–vii.

Jäger, J and H L Ferguson, eds (1991), *Climate Change: Science Impacts and Policy. Proceedings of the Second World Climate Conference*, Cambridge: Cambridge University Press.

Johansson, T B and J N Swisher (1994), Perspectives on 'Bottom-Up' Analyses of the Costs of CO_2 Emission Reductions, in OECD/IEA, *The Economics of Climate Change*, Paris: OECD.

Jones, T (1994), Operational Criteria for Joint Implementation, in OECD/IEA, *The Economics of Climate Change*, Paris: OECD.

Jorgenson, D W and P J Wilcoxen (1990), *Global Change, Energy Prices, and US Economic Growth*, HIER Discussion Paper No 1511, Harvard Institute of Economic Research, Cambridge, MA.

Jorgenson, D W and P J Wilcoxen (1992), *Reducing US Carbon Dioxide Emissions: An Assessment of Different Instruments*, HIER Discussion Paper No 1590, Harvard Institute of Economic Research, Cambridge, MA.

Kalkstein, L S (1989), The Impact of CO_2 and Trace Gas-Induced Climate Changes upon Human Mortality, in J B Smith and D A Tirpak (eds), *The Potential Effects of Global Climate Change on the United States. Appendix G: Health*, Washington DC: EPA.

Kane, S, J Reilly, and J Tobey (1992), An Empirical Study of the Economic Effects of Climate Change on World Agriculture, in *Climatic Change* 21: 17–35.

Karadeloglou, P (1992), Energy Tax versus Carbon Tax: A Quantitative Macro-economic Analysis with the HERMES/Midas Models, in Commission of the European Communities, *The Economics of Limiting CO_2 Emissions*, Special Edition No 1, Luxembourg.

Kaufmann, R, H Y Li, P Pauly and L Thompson (1991), *Global Macroeconomic Effects on Carbon Taxes: A Feasibility Study*, Discussion Paper, University of Toronto, July.

Kokoski, M F and V K Smith (1987), A General Equilibrium Analysis of Partial-Equilibrium Welfare Measures: The Case of Climate Change, in *American Economic Review* 77(3): 331–341.

Kolstad, C D (1991), *Regulating a Stock Externality under Uncertainty with Learning*, Mimeo, Department of Economics, University of Illinois, Urbana.

Kolstad, C D (1993), Looking vs Leaping: the Timing of CO_2 Control in the Face of Uncertainty and Learning, in Y Kaya, N Nakićenović, W D Nordhaus and F Toth (eds), *Costs, Impacts and Benefits of CO_2 Mitigation*, IIASA Collaborative Paper Series, CP93–2, Laxenburg, Austria.

Krause, F, W Bach, and J Kooney (1989), *Energy Policy in the Greenhouse*, London: Earthscan.

Kverndokk, S (1992), *Tradeable CO_2 Emission Permits: The Distribution as an Applied Justice Problem*, Global Environmental Change Working Paper GEC 92–35, Centre for Social and Economic Research on the Global Environment, University College London and University of East Anglia, Norwich.

Kverndokk, S (1993), Global CO_2 Agreements: A Cost Effective Approach, in *Energy Journal* 14(2): 91–112.

Kverndokk, S (1994), *Depletion of Fossil Fuels and the Impacts of Global Warming*, Research Department Discussion Paper No 107, Statistics Norway, Oslo.

Larsen, B and A Shah (1992), *World Fossil Fuel Subsidies and Global Carbon Emissions*, Background Paper for the World Development Report, The World Bank, Washington, DC.

Lashof, D A and D R Ahuja (1990), Relative Contributions of Greenhouse Gas Emissions to Global Warming, in *Nature* 344 (5 April): 529–31.

Leary, N A and J D Scheraga (1993), Lessons for the Implementation of Policies to Mitigate Carbon Dioxide Emissions, in D W Gaskin and J P Weyant (eds), *Reducing*

Carbon Emissions from the Energy Sector: Costs and Policy Options, Cambridge, MA: MIT Press.

Lempert, R J, M E Schlesinger, and J K Hammitt (1994), The Impact of Potential Abrupt Climate Changes on Near-Term Policy Choices, in *Climatic Change* 26(4): 351–76.

Lind, R C, ed (1982), *Discounting for Time and Risk in Energy Policy*, Washington, DC: Resources for the Future.

Lockwood, B (1992), *The Social Costs of Electricity Generation*, Global Environmental Change Working Paper GEC 92–09, Centre for Social and Economic Research on the Global Environment, University of East Anglia, Norwich, and University College London.

Lunde, L (1991), Science and Politics in the Greenhouse. How Robust is the IPCC Consensus?, in *International Challenges. The Fridtjof Nansen Institute Journal*, 11(1): 48–57.

Machina, M (1982), 'Expected Utility' Analysis without the Independence Axiom, in *Econometrica* 50(2): 277–323.

Maddison, D (1993), *The Shadow Price of Greenhouse Gases and Aerosols*, Mimeo, Centre for Social and Economic Research on the Global Environment, University College London and University of East Anglia, Norwich.

Maier-Reimer, E and K Hasselmann (1987), Transport and Storage of CO_2 in the Ocean. An Inorganic Ocean-Circulation Carbon Cycle Model, in *Climate Dynamics* 11(2): 63–90.

Mäler, K G (1989), *Sustainable Development*, Mimeo, Stockholm School of Economics, Stockholm.

Manne, A S (1994), International Trade: The Impacts of Unilateral Carbon Emission Limits, in OECD/IEA, *The Economics of Climate Change*, Paris: OECD.

Manne, A S, R Mendelsohn, and R G Richels (1993), *MERGE – A Model for Evaluating Regional and Global Effects of Greenhouse Gas Reduction Policies*, Mimeo, Electric Power Research Institute, Palo Alto, CA.

Manne, A S and R G Richels (1991), Buying Greenhouse Insurance, in *Energy Policy* 19(6): 543–52.

Manne, A S and R G Richels (1992), *Buying Greenhouse Insurance*, Cambridge MA: MIT Press.

Manne, A S and R G Richels (1993), The EC Proposal for Combining Carbon and Energy Taxes – The Implications for Future CO_2 Emissions, in *Energy Policy* 21(1): 5–12.

Manne, A S and R G Richels (1994a), The Costs of Stabilizing Global CO_2 Emissions: A Probabilistic Analysis Based on Expert Judgments, in *Energy Journal* 15(1): 31–56.

Manne, A S and R G Richels (1994b), CO_2 Hedging Strategies: The Impact of Uncertainty Upon Emissions, in OECD/IEA, *The Economics of Climate Change*, Paris: OECD.

Manne, A S and T F Rutherford (1993), International Trade in Oil, Gas and Carbon Emission Rights – An Intertemporal Equilibrium Model, in Y Kaya, N Nakićenović, W D Nordhaus and F Toth (eds), *Costs, Impacts and Benefits of CO_2 Mitigation*, IIASA Collaborative Paper Series, CP93–2, Laxenburg, Austria.

Manne, A S and C O Wene (1992), *MARKAL-MACRO: A Linked Model for Energy Economy Analysis*, Working Paper BNL–47161, Brookhaven National Laboratory, Upton, NY.

Marglin, S A (1963a) The Opportunity Cost of Public Investment, in *Quarterly Journal of Economics* 77: 274–289.

Marglin, S A (1963b) The Social Rate of Discount and the Optimal Rate of Investment, in *Quarterly Journal of Economics* 77: 95–111.

Markandya, A and D W Pearce (1991), Development, the Environment and the Social Rate of Discount, in *The World Bank Observer* 6(2): 137–152.

Martin, J P, J M Burniaux, G Nicoletti, and J Oliveira-Martins (1992), The Costs of International Agreements to Reduce CO_2 Emissions: Evidence from GREEN, in *OECD Economic Studies* No. 19/Winter.

McKibbin, W J and P J Wilcoxen (1992a), *G-Cubed: A Dynamic Multi-Sector General Equilibrium Growth Model of the Global Economy*, Brookings Discussion Papers No 98, The Brookings Institution, Washington, DC.

McKibbin, W J. and P J Wilcoxen (1992b), *The Global Costs of Policies to Reduce Greenhouse Gas Emissions*, Brookings Discussion Papers No 97, The Brookings Institution, Washington, DC.

Mearns, L O, R W Katz and S H Schneider (1984), Extreme High Temperature Events: Changes in their Probabilities with Changes in Mean Temperature, in *Journal of Climate and Applied Meteorology* 23: 1601–13.

Mendelsohn, R, W Nordhaus, and D Shaw (1992), *The Impact of Climate on Agriculture: A Ricardian Approach*, Discussion Paper No 1010, Cowles Foundation, New Haven, Conn.

Milliman, J D, J M Broadus, and F. Gable (1989), Environmental and Economic Implications of Rising Sea Level and Subsiding Deltas: The Nile and Bengal Examples, in *Ambio* 18(6): 340–45.

Mills, E, D Wilson, and T B Johansson (1991), Getting Started. No Regrets Strategies for Reducing Greenhouse Gas Emissions, in *Energy Policy* 19(6): 526–42.

Morgenstern, R D (1991), Towards a Comprehensive Approach to Global Climate Change Mitigation, in *American Economic Review, Papers and Proceedings* 81(2): 140–45.

Nordhaus, W D (1991a), The Costs of Slowing Climate Change: A Survey, in *Energy Journal* 12(1): 37–65.

Nordhaus, W D (1991b), A Sketch of the Economics of the Greenhouse Effect, in *American Economic Review, Papers and Proceedings* 81(2): 146–50.

Nordhaus, W D (1991c), To Slow or not to Slow: The Economics of the Greenhouse Effect, in *Economic Journal* 101(407): 920–37.

Nordhaus, W D (1991d), Economic Approaches to Greenhouse Warming, in R Dornbusch and J M Poterba (eds), *Global Warming: Economic Policy Responses*, Cambridge, MA: MIT Press.

Nordhaus, W D (1992), *The DICE Model: Background and Structure of a Dynamic Integrated Climate Economy Model of the Economics of Global Warming*, Discussion Paper No 1009, Cowles Foundation, New Haven, Conn.

Nordhaus, W D (1993a), Optimal Greenhouse Gas Reductions and Tax Policy in the 'DICE' Model, in *American Economic Review, Papers and Proceedings* 83(2): 313–17.

Nordhaus, W D (1993b), Rolling the 'DICE': An Optimal Transition Path for Controlling Greenhouse Gases, in *Resources and Energy Economics* 15(1): 27–50.

Nordhaus, W D (1993c), Pondering Greenhouse Policy – Response, in *Science* 259(5 March): 1383–4.

Nordhaus, W D (1993d), *Survey on Uncertainties Associated with Potential Climate Change*, Mimeo, Yale University, New Haven, Conn.

Nordhaus, W D (1994), Expert Opinion on Climate Change, in *American Scientist*, January–February: 45–51.

Norgaard, R B and R B Howarth (1991), Sustainability and Discounting the Future, in R. Costanza (ed), *Ecological Economics*, New York: Columbia University Press.

O'Riordan, T (1992), *The Precaution Principle in Environmental Management*, Global Environmental Change Working Paper GEC 92–03, Centre for Social and Economic Research on the Global Environment, University of East Anglia, Norwich, and University College London.

Ogawa, Y (1991), Economic Activity and the Greenhouse Effect, in *Energy Journal* 12(1): 23–35.

Okken, P A, J R Ybema, D Gerbers, T Kram, and P Lako (1991), *The Challenge of Drastic CO_2 Reduction. Opportunities for New Energy Technologies to Reduce CO_2 Emissions in the Netherlands Energy System up to 2020*, Dutch Energy Research Foundation, Report ECN-C-91-009, Petten, The Netherlands.

Organisation for Economic Cooperation and Development (1991), *OECD Environmental Data Compendium 1991*, Paris: OECD.

Organisation for Economic Cooperation and Development (1992a), *Climate Change. Designing a Practical Tax System*, Paris: OECD.

Organisation for Economic Cooperation and Development (1992b), *Climate Change. Designing a Tradeable Permit System*, Paris: OECD.

Organisation for Economic Cooperation and Development (1992c), *The Economic Costs of Reducing CO₂ Emissions*, OECD Economic Studies, Special Issue, No 19/ Winter, Paris: OECD.

Organisation for Economic Cooperation and Development (1993), *The Costs of Cutting Carbon Emissions: Results from Global Models*, Paris: OECD.

Oliveira-Martins, J, J M Burniaux, and J P Martin (1992), Trade and Effectiveness of Unilateral CO₂ Abatement Policies: Evidence from GREEN, in *OECD Economic Studies* No 19/Winter.

PACE (1990), *Environmental Costs of Electricity*, Pace University Centre for Environmental and Legal Studies, New York: Oceana Publications.

Pachauri, R K and P Bhandari, eds (1992), *Global Warming. Collaborative Study on Strategies to Limit CO₂ Emissions in Asia and Brazil*, New Delhi: Tata McGraw Hill and Asian Energy Institute.

Pachauri, R K and M Damodaoran (1992), 'Wait and See' versus 'No Regrets': Comparing the Costs of Economic Strategies, in I M Mintzer (ed), *Confronting Climate Change, Risks, Implications and Responses*, Cambridge: Cambridge University Press.

Parfit, D (1983), Energy Policy and the Further Future: The Social Discount Rate, in P G Brown and D MacLean (eds), *Energy and the Future*, Totowa, NJ: Rowman and Littlefield.

Parry, I W H (1993), Some Estimates of the Insurance Value against Climate Change from Reducing Greenhouse Gas Emissions, in *Resource and Energy Economics* 15(1): 99–116.

Parry, M (1993), Climate Change and the Future of Agriculture, in *International Journal of Environment and Pollution* 3(1–3): 13–30.

Parry, M L, T R Carter, and N T Konijn, eds (1988), *The Impacts of Climate Variations on Agriculture, 2 Volumes*, Dordrecht: Kluwer.

Pearce, D W (1980), The Social Incidence of Environmental Costs and Benefits, in T O'Riordan and R K Turner (eds), *Progress in Resource Management and Environmental Planning*, Vol 2, Chichester: Wiley.

Pearce, D W (1991), The Role of Carbon Taxes in Adjusting to Global Warming, in *Economic Journal* 101(407): 938–48.

Pearce, D W (1992), *The Secondary Benefits of Greenhouse Gas Control*, Global Environmental Change Working Paper GEC 92–12, Centre for Social and Economic Research on the Global Environment, University College London and University of East Anglia, Norwich.

Pearce, D W (1993a), *Economic Values and the Natural World*, London: Earthscan.

Pearce, D W (1993b), *The Economics of Involuntary Resettlement*, A Report to the World Bank, Centre for Social and Economic Research on the Global Environment, University College London and University of East Anglia, Norwich.

Pearce, D W (1994a), Sustainable Development, in D W Pearce, *Ecological Economics: Essays in the Theory and Practice of Environmental Economics*, London: Edward Elgar, forthcoming.

Pearce, D W (1994b), *Costing the Environmental Damage from Energy Production*, Mimeo, Centre for Social and Economic Research on the Global Environment, University College London and University of East Anglia, Norwich.

Pearce, D W, C Bann and S Georgiou (1992), *The Social Cost of Fuel Cycles*, London: HMSO.

Pearce, D W and S Fankhauser (1993), *Cost Effectiveness and Cost-Benefit in the Control of Greenhouse Gas Emissions*, Paper presented at the Meeting of Working Group 3 of the Intergovernmental Panel on Climate Change, May 1993, Montreal.

Pearce, D W and C A Nash (1981), *The Social Appraisal of Projects. A Text in Cost-Benefit Analysis*, London: Macmillan.

Pearce, D W, T S Swanson, A McGuire, and J Richardson (1991), *Economics, Environment and Health.* A Report for the European Regional Office on the World Health Organisation, Centre for Social and Economic Research on the Global Environment, University College London and University of East Anglia, Norwich.

Pearce, D W and R K Turner (1990), *Economics of Natural Resources and the Environment*, London: Harvester Wheatsheaf.

Pearce, D W and D Ulph (1994), *Discounting and the Early Deep Disposal of Radioactive Waste*, A Report to United Kingdom NIREX Ltd, Centre for Social and Economic Research on the Global Environment, University College London and University of East Anglia, Norwich.

Pearson, M (1992), Equity Issues and Carbon Taxes, in OECD, *Climate Change. Designing a Practical Tax System*, Paris: OECD.

Peck, S C and T J Teisberg (1992), CETA: A Model for Carbon Emissions Trajectory Assessment, in *Energy Journal* 13(1): 55–77.

Peck, S C and T J Teisberg (1993a), CO_2 Emissions Control: Comparing Policy Instruments, in *Energy Policy* 21(3): 222–30.

Peck, S C and T J Teisberg (1993b), Global Warming Uncertainties and the Value of Information: An Analysis Using CETA, in *Resource and Energy Economics* 15(1): 71–97.

Peck, S C and T J Teisberg (1994), *Note on Optimal CO_2 Control Policy with Temperature Dependent Disaster Probabilities*, Mimeo, Electric Power Research Institute, Palo Alto, CA.

Perroni, C and T F Rutherford (1991), *International Trade in Carbon Emission Rights and Basic Materials: General Equilibrium Calculations for 2020*, Mimeo, Wilfrid Laurier University and University of Western Ontario, London, Canada.

Peters, R I and T E Lovejoy, eds (1992), *Global Warming and Biological Diversity*, New Haven: Yale University Press.

Pezzey, J (1992a), Analysis of Unilateral CO_2 Control in the European Community and OECD, in *Energy Journal* 13(3): 159–71.

Pezzey, J (1992b), *Impacts of Greenhouse Gas Control Strategies on UK Competitiveness*, A Report to the UK Department of Trade and Industry, London: HMSO.

Pezzey, J (1992c), *Some Interactions Between Environmental Policy and Public Finance*, Paper presented at the 3rd Annual EAERE Conference, June 1992, Krakow, Poland.

Pigou, A C (1932), *The Economics of Welfare*, 4th ed, London: Macmillan.

Poterba, J M (1991), Tax Policy to Combat Global Warming: On Designing a Carbon Tax, in R Dornbusch and J M Poterba (eds), *Global Warming: Economic Policy Responses*, Cambridge, MA: MIT Press.

Proops, J L R, M Faber, and G Wagenhals (1993), *Reducing CO_2 Emissions. A Comparative Input-Output Study for Germany and the UK*, Berlin: Springer.

Ramsey, F P (1928), A Mathematical Theory of Saving, in *Economic Journal* 138(152): 543–59.

Reilly, J M (1992), Climate-Change Damage and the Trace-Gas-Index Issue, in J M Reilly and M Anderson (eds), *Economic Issues in Global Climate Change*, Boulder, CO: Westview.

Richels, R and J Edmonds (1993), *The Economics of Stabilizing Atmospheric CO_2 Concentrations*, Mimeo, Electric Power Research Institute, Palo Alto, CA.

Rijkswaterstaat (1991), *Rising Waters. Impacts of the Greenhouse Effect for the Netherlands*, Dutch Ministry of Transport and Public Works, The Hague.

Rijsberman, F (1991), Potential Costs of Adapting to Sea Level Rise in OECD Countries, in OECD, *Responding to Climate Change: Selected Economic Issues*, Paris: OECD.

Rose, A (1990), Reducing Conflict in Global Warming Policy. The Potential of Equity as a Unifying Principle, in *Energy Policy* 18(10): 927–35.

Rose, A (1992), Equity Considerations of Tradeable Carbon Emission Entitlements, in UNCTAD, *Combating Global Warming. Study on a Global System of Tradeable Carbon Emission Entitlements*, Geneva: UNCTAD.

Rosebrock, J (1993), *Time-Weighting Emission Reductions for Global Warming Projects – A Comparison of Shadow Price and Emission Discounting Approaches*, Mimeo, The World Bank, Washington, DC.

Rosenberg, N J and P R Crosson (1991), The MINK Project: A New Methodology for Identifying Regional Influences of, and Responses to, Increasing Atmospheric CO_2 and Climate Change, in *Environmental Conservation* 18(4): 313–22.

Rosenzweig, C and M L Parry (1994), Potential Impact of Climate Change on World Food Supply, in *Nature* 367(13 January): 133–8.

Rosenzweig, C, M Parry, K Frohberg, and G Fisher (1993), *Climate Change and World Food Supply*, Environmental Change Unit, Oxford.

Rotmans, J, H de Boois, and R Swart (1990), An Integrated Model for the Assessment of the Greenhouse Effect: The Dutch Approach, in *Climatic Change* 16: 331–5.

Rutherford, T F (1993), The Welfare Effects of Fossil Carbon Restrictions: Results from a Recursively Dynamic Trade Model, in OECD, *The Costs of Cutting Carbon Emissions: Results from Global Models*, Paris: OECD.

Schelling, T (1992), Some Economics of Global Warming, in *American Economic Review* 82(1): 1–14.

Scheraga, J D and N A Leary (1992), Improving the Efficiency of Policies to Reduce CO_2 Emissions, in *Energy Policy* 20(5): 394–404.

Scheraga, J D and N A Leary (1994), Costs and Side Benefits of Using Energy Taxes to Mitigate Global Climate Change, in *Proceedings of the 86th Annual Conference*, National Tax Association, Washington, DC.

Scheraga, J D., N A Leary, R J Goettle, D W Jorgenson, and P J Wilcoxen (1993), Macroeconomic Modelling and the Assessment of Climate Change Impacts, in Y Kaya, N. Nakićenović, W D Nordhaus and F Toth (eds), *Costs, Impacts and Benefits of CO_2 Mitigation*, IIASA Collaborative Paper Series, CP93–2, Laxenburg, Austria.

Schmalensee, R (1993) Comparing Greenhouse Gases for Policy Purposes, in *Energy Journal* 14(1): 245–255.

Schneider, S H (1993), Pondering Greenhouse Policy, in *Science* 259(5 March): 1381.

Schneider, S H, P H Gleick and L O Mearns (1990), Prospects of Climate Change, in P E Waggoner (ed.), *Climate Change and US Water Resources*, New York: Wiley.

Schneider, S H and S L Thompson (1981), Atmospheric CO_2 and Climate: Importance of the Transient Response, in *Journal of Geophysical Research* 86 (20 April): 3135–47.

Sedjo, R A and A M Solomon (1989), Climate and Forests, in N J Rosenberg, W E Easterling III, P R Crosson, and J Darmstadter (eds), *Greenhouse Warming: Abatement and Adaptation*, Washington, DC: Resources for the Future.

Shackleton, R, M Shelby, A Cristofaro, R Brinner, J Yanchar, L Goulder, D Jorgenson, P Wilcoxen, P Pauly, and R Kaufmann (1992), *The Efficiency Value of Carbon Tax Revenues*, Paper submitted to the Stanford Energy Modelling Forum Report 12 (EMF 12), US Environmental Protection Agency, Washington, DC.

Shah, A and B Larsen (1992), *Carbon Taxes, the Greenhouse Effect and Developing Countries*, Background Paper No 6, World Development Report 1992, The World Bank, Washington, DC.

Sinclair, P (1992), High does Nothing and Rising is Worse: Carbon Taxes Should be Kept Declining to Cut Harmful Emissions, in *Manchester School*, 60: 41–52.

Smith, J B and D A Tirpak, eds (1989), *The Potential Effects of Global Climate Change on the United States*, Washington, DC: US Environmental Protection Agency.

Smith, K (1992), *Environmental Hazards. Assessing Risk and Reducing Disaster*, London: Routledge.

Smith, S (1992a), The Distributional Consequences of Taxes on Energy and the Carbon Content of Fuels, in Commission of the European Communities, *The Economics of Limiting CO$_2$ Emissions*, Special Edition No 1, Luxembourg.

Smith, S (1992b), Taxation and the Environment: A Survey, in *Fiscal Studies* 13(4): 21–57.

Smith, S (1994), Who Pays for Climate Change Policies? Distributional Side-Effects and Policy Responses, in OECD/IEA, *The Economics of Climate Change*, Paris: OECD.

Solow, R (1974), The Economics of Resources or the Resources of Economics, in *American Economic Review* 64(2): 1–14.

Solow, R (1992), *An Almost Practical Step Toward Sustainability*, An Invited Lecture on the Occasion of the Fortieth Anniversary of Resources for the Future, RFF, Washington, DC.

Standaert, S (1992), The Macro-Sectoral Effects of an EC-Wide Energy Tax: Simulation Experiments for 1993–2005, in Commission of the European Communities, *The Economics of Limiting CO$_2$ Emissions*, Special edition No 1, Luxembourg.

Subak, S and M Kelly (1993), *Projected Warming under Alternative Commitments*, A Review Draft for INC 9, Centre for Social and Economic Research on the Global Environment, University of East Anglia, Norwich and University College London.

Swart, R J and P Vellinga (1994), The 'Ultimate Objective' of the Framework Convention on Climate Change Requires a New Approach in Climate Change Research. An Editorial, in *Climatic Change* 26(4): 343–50.

Symons, E J, J R L Proops, and P W Gay (1994), Carbon Taxes, Consumer Demand and Carbon Dioxide Emissions: A Simulation Analysis for the UK, in *Fiscal Studies* 15(2): 19–43.

Tahvonen, O (1993), *Carbon Dioxide Abatement as a Differential Game*, Discussion Papers in Economics and Business Studies No 4, University of Oulu, Finland.

Titus, J (1992), The Cost of Climate Change to the United States, in S K Majumdar, L S Kalkstein, B Yarnal, E W Miller, and L M Rosenfeld (eds), *Global Climate Change: Implications, Challenges and Mitigation Measures*, Pennsylvania: Pennsylvania Academy of Science.

Titus, J G, R A Park, S P Leatherman, J R Weggel, M S Greene, P W Mausel, S Brown, C Gaunt, M Trehan, and G Yohe (1991), Greenhouse Effect and Sea Level Rise: The Cost of Holding Back the Sea, in *Coastal Zone Management* 19: 172–204.

Tol, R S J (1993), *The Climate Fund. Survey of Literature on Costs and Benefits*, Working Document W 93/01, Free University of Amsterdam (revised).

Tol, R S J (1994), The Damage Costs of Climate Change. A Note on Tangibles and Intangibles, Applied to DICE, in *Energy Policy* 22(5): 436–8.

Turner, R K (1991), Valuation of Wetland Ecosystems, in D W Pearce and H Opschoor (eds), *Persistent Pollutants: Economics and Policy*, Dordrecht: Kluwer.

Turner, R K, P Doktor, and N Adger (1994a), Assessing the Costs of Sea Level Rise: East Anglian Case Study, in *Environment and Planning A*, forthcoming.

Turner, R K, S. Subak, N Adger and J Parfitt (1994b), *How are Socio-Economic Systems Affected by Climate Related-Changes in the Coastal Zone?*, A Report to IPCC, Working Group 2, Subgroup B, Centre for Social and Economic Research on the Global Environment (CSERGE), University of East Anglia and University College London.

Turner, R K and T Jones, eds (1990), *Wetlands. Market & Intervention Failures. Four Case Studies*, London: Earthscan.

UK Climate Change Impact Review Group (1991), *The Potential Effects of Climate Change in the United Kingdom*, Report prepared at the Request of the Department of the Environment, London: HMSO.

Ulph, A, D Ulph, and J Pezzey (1991), *Should a Carbon Tax Rise or Fall over Time?*, Discussion Paper No 9115, Department of Economics, University of Southampton.

Ulph, A and D Ulph (1994a), *Global Warming, Irreversibility and Learning: Some Clarificatory Notes*, Mimeo, Centre for Social and Economic Research for the Global Environment, University College London and University of East Anglia, Norwich.

Ulph, A and D Ulph (1994b), *Who Gains from Learning About Global Warming?* Department of Economics Working Paper No 94–11, University of Birmingham.

Ulph, D (1992), *A Note on the 'Double Benefit' of Pollution Taxes*, Discussion Paper No 92/317, Department of Economics, University of Bristol.

UNCTAD (1992), *Combating Global Warming. Study on a Global System of Tradeable Carbon Emission Entitlements*, Geneva: UNCTAD.

UNEP (1992), *UNEP Greenhouse Gas Abatement Costing Studies. Phase One Report*, UNEP Collaborating Centre on Energy and Environment and Risø National Laboratory, Roskilde, Denmark.

United Nations (1990), *National Accounts Statistics: Main Aggregates and Detailed Tables, 1988*, New York.

United Nations (1991), *Demographic Yearbook 1989*, New York.

Viscusi, W K (1993), The Value of Risks to Life and Health, in *Journal of Economic Literature*, XXXI(December): 1912–46.

von Weizsäcker, E U and J Jesinghaus (1992), Oekologische Steuerreform. Teil I: Europische Ebene, in S P Mauch, R Iten, E U von Weizsäcker, and J Jesinghaus, *Oekologische Steuerreform. Europäische Ebene und Fallbeispiel Schweiz*, Zürich: Rüegger.

Vouyoukas, L (1993), The IEA Medium Term Energy Model, in OECD, *The Costs of Cutting Carbon Emissions: Results from Global Models*, Paris: OECD.

Waggoner, P E (1990), *Climate Change and US Water Resources*, New York: Wiley.

Weihe, W H and R Mertens (1991), Human Well-Being: Diseases and Climate, in J Jäger and H L Ferguson (eds), *Climate Change: Science, Impacts and Policy*. Proceedings of the Second World Climate Conference, Cambridge: Cambridge University Press.

Weitzman, M L (1974), Prices vs. Quantities, in *Review of Economic Studies* 41(4): 477–91.

Weitzman, M (1994), On the 'Environmental' Discount Rate, in *Journal of Environmental Economics and Management* 26(2): 200–209.

Wene, C O (1993), Top Down – Bottom Up: A Systems Engineers's View, in Y Kaya, N Nakićenović, W D Nordhaus and F Toth (eds), *Costs, Impacts and Benefits of CO_2 Mitigation*, IIASA Collaborative Paper Series, CP93-2, Laxenburg, Austria.

Weyant, J P (1993), Costs of Reducing Global Carbon Emissions, in *Journal of Economic Perspectives* 7(4): 27–46.

Whalley, J and R Wigle (1991a), Cutting CO_2 Emissions: The Effects of Alternative Policy Approaches, in *Energy Journal* 12(1): 109–24.

Whalley, J and R Wigle (1991b), International Incidence of Carbon Taxes, in R Dornbusch and J M Poterba (eds), *Global Warming: Economic Policy Responses*, Cambridge, MA: MIT Press.

Whalley, J and R Wigle (1993), Results for the OECD Comparative Modelling Project from the Whalley-Wigle Model, in OECD, *The Costs of Cutting Carbon Emissions: Results from Global Models*, Paris: OECD.

Wigley, T M L, and S C B Raper (1992), Implications for Climate and Sea Level of revised IPCC Emissions Scenarios, in *Nature* 357: 293–300.

Williams, R H (1990), Low-Cost Strategies for Coping with CO_2 Emission Limits (A Critique of 'CO_2 Emission Limits: An Economic Cost Analysis for the USA' by Alan Manne and Richard Richels), in *Energy Journal* 11(3): 35–59.

Wilson, D and J Swisher (1993), Exploring the Gap: Top-Down vs Bottom-Up Analyses of the Cost of Mitigating Global Warming, in *Energy Policy* 21(3): 249–63.

Wirth, D A and D A Lashof (1991), Beyond Vienna and Montreal – Multilateral Agreements on Greenhouse Gases, in *Ambio* 19: 305–310.

World Bank (1990), *The World Development Report 1990*, Oxford: Oxford University Press.

World Bank (1992), *World Development Report 1992*, Oxford: Oxford University Press.

World Health Organisation (1990), *Potential Health Effects of Climatic Change*, Geneva: WHO.

World Resources Institute (1990), *World Resources 1990–91: A Guide to the Global Environment*, Vol 4. World Resources Series. New York: Oxford University Press.

World Resources Institute (1992), *World Resources 1992–93: A Guide to the Global Environment*, Vol 5. World Resources Series. New York: Oxford University Press.

Yohe, G W (1991), Uncertainty, Climate Change and the Economic Value of Information: An Economic Methodology for Evaluating the Timing and Relative Efficacy of Alternative Response to Climate Change with Application to Protecting Developed Property from Greenhouse Induced Sea Level Rise, in *Policy Sciences* 24: 245–69.

Index